ALAMANCE COMMUNITY COLLEGE
P.O. BOX 8000
GRAHAM, NC 27253-8000

ALAMANCE COMMUNITY COLLEGE
P.O. BOX 8000
GRAHAM, NC 27253-8000

ALSO BY LISA ENDLICH

Goldman Sachs: The Culture of Success

SIMON & SCHUSTER

New York London

Toronto Sydney

OPTICAL ILLUSIONS

LUCENT AND THE CRASH OF TELECOM

LISA ENDLICH

SIMON & SCHUSTER
Rockefeller Center
1230 Avenue of the Americas
New York, NY 10020

For information about special discounts for bulk purchases,
please contact Simon & Schuster Special Sales at
1-800-456-6798 or business@simonandschuster.com

Designed by Jeanette Olender
Manufactured in the United States of America

1   3   5   7   9   10   8   6   4   2

Library of Congress Cataloging-in-Publication Data
Endlich, Lisa, date.
Optical illusions : Lucent and the crash of Telecom / Lisa Endlich.
p.   cm.
Includes bibliographical references and index.
1. Lucent Technologies (Firm)—History.   2. Telecommunication
equipment industry—United States—History.   3. Data transmission
equipment industry—United States—History.   I. Title.
HD9696.T444L834   2004
338.7'62382'0973—dc22   2004052115
ISBN 0-7432-2667-4

ACKNOWLEDGMENTS

There is no way this account of Lucent could have shed any light on its subject had not dozens and dozens of people, who spent many years in Murray Hill, offered their time and insights. Most of these generous souls have asked that I leave their names off this manuscript, and so I simply offer my thanks.

In particular I would like to offer my appreciation for their time and patience to: Bill Brinkman, Robert Feidelson, Curtis Crawford, Lance Boxer, Jeong Kim, David Nadler, Dan Plunkett, Jim Lusk, Gary Nilsson, Roger Boyce, Hilary Mine, Bob Hewitt, William Holstein, Marc Schweig, Anthony Langham, Rich Moscioni, Steve Toll, Steven Russell, Steve Akers, Chip Leveroni, Dawn Truax, Mark Mortenson, Alan Turmillo, Maddie Carter, and Mike Hendren.

Bob Bender guided me with supreme editorial skill, showing insight and precision at every turn. He and Johanna Li led me through the baffling maze of publishing.

Geri Thoma read and reread proposals, steering me to the right publisher.

Steve Levy was both dramatis persona and masterful critic, but perhaps most important, he made me laugh.

Jeffrey Williams led me through the IPO in such careful detail that I felt as though I had been there.

The good fortune of having your oldest friend become a writer and editor should not be underestimated, and again I proffer Michelle Miller-Adams my gratitude.

I only hope that in whipping up the occasional tuna noodle casserole that I can offer my thanks to a husband who, while insisting that I had become a troglodyte, was happy to take on many of my parental duties with our sons. Perhaps the greatest gift he has given me, in this endeavor and all others, is that he kept his sense of humor and balance long after I had dispensed with my own. There are no words to express my appreciation for these gifts.

FOR MARK

CONTENTS

Preface   1

CHAPTER **ONE**   **LOOKING BACK**   5

CHAPTER **TWO**   **EARLY YEARS**   9

CHAPTER **THREE**   **McGINN AND SCHACHT**   21

CHAPTER **FOUR**   **1996**   39

CHAPTER **FIVE**   **1997**   77

CHAPTER **SIX**   **1998**   93

CHAPTER **SEVEN**   **1999**   133

CHAPTER **EIGHT**   **2000**   179

CHAPTER **NINE**   **2001**   225

CHAPTER **TEN**   **2002–2003**   255

Notes   267

Index   289

OPTICAL ILLUSIONS

PREFACE

Henry Schacht had simply wanted to issue a press release, a short notice saying that as of October 6, 1997, he was stepping down as CEO of Lucent Technologies, to be replaced by his friend and partner, the company's president, Rich McGinn. Kathy Fitzgerald, Lucent's senior vice president of public relations, respectfully disagreed. From her three decades in public relations at AT&T, she knew how to downplay bad news and grab as much press coverage as she could when she had a good story to report. She convinced Schacht that an all-employee broadcast was needed and set about putting the pieces in place. Fitzgerald focused on the message, but she was presenting a picture as well as a story. A good presentation took time, and that is precisely what her bosses, Schacht and McGinn, had failed to give. No matter; in the few hours between the time they informed her of the leadership transition and were ready to announce it to the world, she secured a large gray auditorium at Lucent's Murray Hill headquarters in New Jersey, covered it with red and white balloons, packed it with an employee audience, arranged a television hookup to every one of the company's 120,000 employees, and ordered a cake. The cake would have to be simple; baking was one of the few things even she, with all her years of experience, could not rush.

Now, only hours after she had put the wheels in motion, Fitzgerald stood before her colleagues and in her soft voice, assured the standing-room-only crowd, "We're not in violation of any fire codes yet, but

1

we're working on it." And then, with a smile, said, "It's just another of the kind of risk we're used to taking at this company, the kind of people we are." As she attempted to read through a brief introduction of Schacht, she was interrupted repeatedly with cheers from her colleagues at every mention of their leader's name. They knew what his message would be. Only moments earlier, Fitzgerald had let the good news slip.

Together, for the past year, Schacht and McGinn had run Lucent, the newly created telecommunications equipment giant. Here was the moment of handoff, as the tall, slim Schacht, an elder statesman in corporate America, stepped aside in favor of his younger president. Their partnership had been created by AT&T chairman Robert Allen at the time he spun off Lucent as a separate business. Allen had paired Schacht, the former AT&T board member and Cummins Engine CEO, with McGinn, the leader of the largest AT&T business that would become Lucent. Schacht brought experience and credibility to the new company; McGinn brought a deep-seated knowledge of telecommunications and a vision for the future. McGinn had been deeply disappointed with Allen's decision. Maybe, he had thought, he would get stuck with a nonexecutive chairman. But it was a truly unpleasant surprise to find Allen had given Schacht the CEO title. Yet, by all reckonings, the duo had been a success; their leadership had been effective and their personal affection was real. Lucent was riding high. Already a Wall Street darling, it was in a stronger financial position than anyone had imagined possible for a dreary equipment business that had spent more than a century as a captive of the phone company. The company's stock had doubled in the thirteen months since it became public, and in the eyes of its employees and investors Schacht and McGinn were heroes.

Schacht, in shirtsleeves and tie, spoke first, beaming at the audience and trying to suppress his laughter. In his flat midwestern accent, he recounted a conversation with a colleague earlier that day in which he had thought he would surprise her with the news of McGinn's ascendancy, saying, "She said we've all known something was up because he [McGinn] started to come to work in suits and socks." Then, eyeing McGinn offstage, he added, "And he just told me that he gets the mike last, so I'm in trouble." Later, McGinn acquitted himself and assured his employees that lest they worry there was a new Rich, nothing had changed; his casual clothes were simply at the cleaner's. In announcing the transition, Schacht described his partnership with McGinn in reverential tones. Such relationships in business, he said, were rare. Reflecting on the previous year and a half, Schacht offered, "Nothing has been more pleasurable, more important and more unique than the partnership Rich and I have had during this period of time." The bond between the two men looked heartfelt as each man tried to deflect the attention from himself, heaping praise upon his partner.

After confiding to his audience that in his business life there had been no greater pleasure than handing the baton off to McGinn, Schacht enumerated the details of the new arrangement. His speech betrayed his characteristic interest in procedures and the finer points. Schacht likes a good detail, and those who have worked with him tell stories of his managing minutiae. Before any important meetings, Schacht likes to know the list of attendees and the configuration of the table. After he has made it clear where the place cards should be laid, he arrives early to double-check that his instructions have been followed and make any last-minute seating changes. On his recommendation, the Lucent board had named McGinn CEO the day before. In turn, McGinn had asked Schacht to stay on as chairman until the following year. The board had made this decision only the night before in a special meeting. This fact might have triggered a great deal of interest if Schacht's audience had known that some of the most senior executives standing in their midst had begged him not to relinquish his role. McGinn was not ready, they argued; give it more time. Nowhere was there any hint of the fact that Schacht had been reluctant to leave his post. No one looking on at that moment could have guessed that he was ceding power based upon a deal the two men had cut a year earlier.

All of this went unsaid as Schacht described the transition as a relay race; Fitzgerald, as is her trademark, had supplied him with a prop, an actual wood baton to hand to McGinn. As in a race, Schacht offered, a new surge of enthusiasm would come with a new runner. When Schacht left the stage, he passed the wooden dowel to McGinn and embraced the younger man.

McGinn, with his dark hair and Harry Potter glasses, may have been Schacht's protégé, but he could not have been more different. Uncomfortable with contrived gestures and genuinely surprised to be handed an actual baton, McGinn stared at the odd piece of wood in his hand for a moment as if figuring out what he could possibly say. McGinn, in his deep, resonant voice, delivers his humor deadpan. In an unintended twist of irony, he told his audience, "I am somewhat concerned here," as he gazed down at the short piece of wood. "It says here 'first annual baton pass.' I don't understand completely what that means, but I guess Henry will explain it later to me."

As usual, McGinn's prepared remarks remained rolled up in his hand as he ad-libbed with the ease of a stand-up comic, expressing his thoughts with humor and warmth. McGinn's expression became more serious when he shared with his audience the conversation he had had with the board the preceding day. Lucent's success and his ascension as CEO were due to what McGinn described as "Schacht's character, qualities and values." He called Schacht Lucent's founding father and noted that with his shock of white hair, he truly looked the part. Neither man chose to make a policy speech, to touch on new initiatives or directions for the company. Neither man even hinted at the possibility that McGinn would set a new course for Lucent.

The transfer of power was billed as a celebration, and Fitzgerald had arranged for a chef to wheel out a large white sheet cake with the words "Congratulations Rich McGinn" frosted across it. Again McGinn stared in obvious discomfort with yet another prop, and still on the stage he plaintively asked Fitzgerald, "Does this mean I have to come down and cut this thing? They'd like that?" Resigned to slicing the enormous confection, he said, "I promise to do my best to avoid the Anthony Perkins–Bates Motel approach to cutting." McGinn and Schacht each sliced a piece of cake; then the chef took over the knife and McGinn's employees swarmed him. McGinn passed easily among them, sharing handshakes, hugs, and pats on the back with an ease and grace that would put any experienced politician to shame. It was the beginning of a new era, and, McGinn told his colleagues, the period of creation was over. McGinn had waited, acceded to the partnership with Schacht, but now the company was his to run. As the two men stood before their employees celebrating their joint accomplishments, who could have guessed that one day Schacht would fire McGinn?

CHAPTER ONE LOOKING BACK

> *The underpinnings of the emerging telecom*
> *bubble were a phenomenal miscalculation. At*
> *the time it seemed like a logical progression of*
> *history: cellular, the Internet, the new thing. It*
> *was bold, it was risky, it was expensive. And it*
> *was wrong.*
>
> David Barden
> telecommunications analyst, Bank of America

The stock market bubble of the late 1990s will undoubtedly be re-membered as the "Internet craze" or "dot-com mania," but this would be selling the madness short. While the fortunes of individual dot-coms have provided journalists and financial historians with some of the most dramatic, colorful, and apocryphal tales, they form only a peripheral portion of a much larger and more significant tale. No industry looked more promising or bled more money than telecommunications. With a collapse in market capitalization of more than $4 trillion and job losses in excess of 500,000 between 1999 and 2002, the telecom meltdown ranks as the greatest stock market debacle ever.

Lucent embodied the transformation in telecommunications. Born in April 1996 as a spin-off of the foundering telecom giant AT&T, Lucent had once been part of the largest regulated monopoly in the United States and thus was protected from the vagaries of the marketplace.

Lucent was the largest initial public offering ever, and because each AT&T shareholder became a Lucent shareholder (along with the new investors attracted by Lucent's stellar performance), it was not long before Lucent was the most widely held stock in America.

Market booms and busts have been studied over time from a macro perspective, with the observer stepping back to examine their insidious effects on the economy as a whole. Perhaps some light can be shed on the boom and bust at the turn of the century by adopting a micro point of view, by studying the path that leads to madness through the actions of a single company that in every way embodied its times. Lucent was born at the inception of the greatest market bubble ever, at the outset of an investment boom that poured billions of dollars into the telecommunications industry. And while understanding this mania in investment is essential to comprehending the company, the reverse may be true as well.

Lucent's startling transformation from an obscure division of AT&T to the sixth-largest corporation in America and then to a company gasping for its financial life left investors and observers by turns delighted, shocked, confused, and finally dismayed as they sought an explanation for the company's descent. The speed of Lucent's decline was as staggering as its magnitude. In the space of twenty-four months, the market capitalization of the company dropped by a quarter of a trillion dollars as its stock price plummeted by 99 percent. Lucent went from spending $100 million to advertise its new name to turning off lights and shuttering bathrooms to save pennies. This is the story of a financially sound company steeped in world-class talent, dominant in one of the world's fastest-growing industries, that in the space of two painful years found itself branded with a junk-bond credit rating, under investigation by the SEC for its fraudulent accounting practices, fighting off rumors of insolvency, and, hat in hand, begging its bankers for a little more time.

Situated on the seam where the new economy met the old, Lucent's prominence alone might have made its downfall a moderately interesting tale. But the story is surprisingly dramatic for a telecom equipment vendor, complete with a boardroom coup, a stock rise and fall in the shape of Mount Everest, and a series of financial and product decisions that left the company enfeebled. This is not a tale that can be charted over decades by pointing to a corporate culture decaying from within or a market in, say, buggy whips or bonnets disappearing in its entirety. Nor is it the story of teenagers popping out of a garage with an invention that would earn them billions, yet finding themselves unable to run the company they built. Lucent's leaders were the men and women of AT&T, the people who practically invented the modern corporate structure. As an old-world stalwart turned new-world bellwether, Lucent experienced the fallout from this historic event with as much force as any ethereal dot-com. When it was over, CEO Henry Schacht was forced to admit that the frenzy that engulfed the stock markets had badly distorted the company's value structure, tampered with its moral compass, and led to decisions that would never have been made in saner times.

The moment of inflection for Lucent came when the U.S. government passed the Telecommunications Act of 1996, igniting competition and forever shattering the rigid confines of this cozy, quasi-monopolistic industry. For Lucent, this regulatory change was a godsend. While Lucent remained inside AT&T, its customers had been AT&T itself and the local Bell phone companies. Soon hundreds of upstart telecom carriers were in a race to build their own state-of-the-art networks, relying almost entirely on borrowed funds. A spending free-for-all broke out, and as the world's largest purveyor of telecommunications equipment, Lucent could not have been better situated to take advantage of this investment binge.

Henry Schacht is known for his business wisdom, and in his earliest days as CEO, he accurately foretold the difficulties that would plague his company and its industry in this unsettled environment: "The pace of change is escalating dramatically at a time when visibility and predictability are declining dramatically." Management would need to pick up the pace, and for those coming from AT&T "the future is getting more opaque. My analogy is that it is the equivalent of a racecar driver accelerating in an increasingly dense fog. This is not a description that human beings find very comforting. We instinctively shy away from rapid destabilizing change; an element of fear is introduced." Even he would be surprised at the prescience of his predictions.

CHAPTER TWO   EARLY YEARS

*When everyone had a telephone then AT&T
had to go off and do something else, and that
was very hard to do.*

Arno Penzias

Nobel Prize winner and former Bell Labs chief

Lucent's birth was announced on September 20, 1995, when Robert Allen, chairman of AT&T, stunned Wall Street, most of his 300,000 employees, and the company's 2.3 million shareholders by announcing the trivestiture of America's largest corporation. "Changes in customer needs, technology and public policy are radically transforming our industry," Allen noted at the time. "We now see this restructuring as the next logical turn in AT&T's journey since divestiture." Although his timing was a surprise, Allen was simply facing up to the radically altered competitive landscape that AT&T now inhabited. AT&T was one of the last large vertically integrated phone systems, capable of inventing, developing, and installing every aspect of telephone service, from transatlantic cables to the Princess phone (available in pink, tan, white, or turquoise) once coveted by every teenage girl in America. Yet the telecom giant had sat and watched the benefits of its monolithic structure slip away.

The story of Lucent's separation from AT&T goes back to the era of Western Electric, once the country's largest manufacturer of consumer electrical goods. Western Electric would later be renamed AT&T Network Systems and in turn would form the core of Lucent's business. Established in 1869, seven years before Alexander Graham Bell patented the telephone, Western Electric was originally the manufacturing arm of Western Union. After the invention of the telephone, Western Electric advertised itself (using quaint, crudely drawn stick men with open circles for heads) as the manufacturer, purchaser, and distributor of 43,000 varieties of telephone apparatus. It became a unit of the Bell System in 1882. With the integration of Western Electric, the growing Bell System strengthened its competitive position. During the first decade of the twentieth century, AT&T acquired dozens of smaller regional carriers, bringing upon itself the scrutiny of the U.S. government, which argued that this structure gave the phone carrier excessive pricing power.

AT&T entered a long and unhappy relationship with the U.S. government. For more than half a century, each accused the other of unfair play, and their differences of opinion tied up the courts for decades. As early as 1910, the Interstate Commerce Commission formally investigated AT&T over its predatory practices toward smaller rivals. The outcome of the investigation was that AT&T and the government reached an agreement under which the phone carrier began life as a federally regulated monopoly. The Federal Communications Commission investigated the structure of the Bell System in the 1930s. The Justice Department then filed suit in 1949 and for seven years attempted to sever Western Electric from AT&T, thereby separating the manufacturing of telephone equipment from the providing of service. The government argued that AT&T had a "captive monopoly" in the phone equipment business, as Western Electric supplied all of the country's telephones, completely shutting out competition and earning itself outsized profits. AT&T countered that vertical integration was in the consumer's interest because it provided economies of scale, uniformity of voice signal, and high engineering standards. In the resulting consent decree of 1956, Western Electric was permitted to remain part of the Bell System on the condition that it limit its manufacturing, although not its research, to telephone equipment and defense work.

In the 1956 decree, the government stipulated that Bell Labs, AT&T's research arm, license all of its patented technological innovations to anyone willing to pay a reasonable royalty, even to potential competitors. Initially, those at Bell Labs did not see this as much of a sacrifice because the commercial potential of many of their discoveries and inventions was entirely unforeseen and commerce itself not highly valued. By making the technology of the Bell System available to one and all, AT&T spawned entire industries based on Bell Labs research from which it was legally barred. In 1956 this had seemed a minor concession; in the end, it may have been the most costly business decision in history.

Had AT&T seized the technological advantages that were already within its grasp instead of fighting so hard to maintain an obsolete monopoly, it might now be the leader in any number of high-tech industries. Bell Labs' mind-boggling list of innovations included the transistor, laser, optical amplification, cellular transmission, frame relay packet switching, UNIX operating system, and the computer language C. All were commercialized by other companies, among them Sony, Ciena, Nortel, Cascade, Sun Microsystems, Hewlett-Packard, and Microsoft. A student of history, McGinn fully appreciated the compromise AT&T had made. "The research that was done years ago, we gave it all away," McGinn recalls. "Same with wireless. We gave that stuff away to the world, and many companies have prospered." In a painful twist of irony, McGinn would later spend billions of dollars acquiring the very companies that had successfully managed to capitalize on Bell Labs' innovations.

By the 1970s, AT&T was the country's largest private employer, with over a million people carrying the AT&T employee card. Every year, as Allen liked to boast, the telephone monopoly bought more than a quarter million telephone poles, 20,000 tons of paper for telephone directories, and 500 million rolls of paper towels. It was a corporate behemoth, the largest on Earth, according to Allen, with assets totaling more than a trillion dollars in today's terms, or more than GM, IBM, GE, U.S. Steel, Kodak, and Xerox combined. Nine out of ten American homes had telephone service, and virtually every one of them was connected through an AT&T-manufactured cable, wire, switching system, and handset. When AT&T said, "Reach out and touch someone," there were few places in America outside its grasp.

In 1974 the Justice Department was again breathing down AT&T's neck at the behest of a poorly capitalized, unknown corporation that was marketing a private phone service between St. Louis and Chicago. Microwave Communications of America, Inc., or MCI, needed the use of AT&T's network and was willing to pay a fee. Threatened by even the smallest inroad into its vast market, AT&T did not cooperate. Face-to-face meetings between AT&T chairman John deButts and MCI chairman William McGowan soon turned nasty and ended with McGowan telling deButts, "I have plenty of money. I can spend it on litigation, or I can spend it on construction. I would prefer to spend it on construction."

"I've heard threats like that before," said deButts. "I won't be coerced."

Like a bad movie that just keeps rerunning, the Justice Department again filed an antitrust suit against AT&T. AT&T tangled with the Justice Department in the courts for eight years, by which time the political tide had turned against it and a new, more flexible CEO was in place. The phone company settled its case. The 1982 consent decree reversed the stipulations in the 1956 decree. Long-distance and local calling services were wrenched apart, and the monolithic monopoly, intractable and highly resistant to change, was vanquished. In its place, competition rose.

AT&T managed to hold on to Western Electric and Bell Labs once again, this time at the cost of setting free the local phone carriers. The company made this deal because the new CEO, Charlie Brown, believed that research and equipment manufacturing would be two of the central elements in AT&T's emergence as a major factor in the "information age." The new decree gave birth to what would fondly be called the Baby Bells or RBOCs (Regional Bell Operating Companies), including NYNEX, Ameritech, Bell Atlantic, Bell South, Pacific Telesis Group, Southwestern Bell, and U.S. West. AT&T was left with three major businesses, its traditionally profitable long-distance service (known as AT&T Long Lines), research (Bell Labs), and manufacturing (Western Electric, renamed Network Systems), along with aspirations to enter the computing business with haste. From the moment of the divestiture, it was apparent that this agreement was also flawed and that AT&T's problems had only begun. The separation of phone service into local and long distance was an artifice that could not endure.

AT&T's attempts to diversify its businesses are a catalog of misadventures. After divestiture, AT&T envisioned taking on computer giant IBM and set about finding an acquisition target that would make this possible. It looked closely at a number of companies, Apple Computer among them. In 1990 McGinn, then president of AT&T's computer systems unit, tried to convince top management to pursue a merger with Hewlett-Packard, but the idea never got a hearing. Then, in 1991, AT&T rushed into a $7.4 billion hostile takeover of NCR (National Cash Register) as the means to challenge IBM. "Everyone knew at the time NCR was a third- or fourth-rate computer company," recalls Michael Porter, a professor at Harvard Business School, "but somehow AT&T thought they could put it together and there'd be all this synergy." Allen discussed NCR with McGinn before the purchase, and the younger executive was very hesitant. Given your aspirations in this marketplace, the company you want to be buying isn't the company you're buying, McGinn warned him.

Allen was confident that NCR would provide AT&T with a solid foundation in the computer business. Yet at no point could this transaction have been described as a success. The takeover fight was so bitter that the CEO of NCR left the day the merger was completed. In short order, AT&T bought NCR, changed its name to AT&T Computer Systems, then lost $2.6 billion competing in a market that it had entered late. Four years later, AT&T changed the entity's named back to NCR and spun it off to shareholders. Over the course of a decade, AT&T entered and then withdrew from the computer, credit card, software, proprietary online service, and Internet content businesses. In the interim, while it pursued this array of computer-related activities, AT&T missed enormous opportunities much closer to home in the data networking business, diminished its share price, and damaged its credentials as a manager of a high-tech company.

AT&T's problems in the post-divestiture period extended to its core businesses. The new, more aggressive phone carriers offered customers bargain calling packages and quickly ate away at AT&T's long-distance market share. In the ten years after divestiture, some eight hundred suppliers of long-distance services sprang up and the price of a long-distance call declined by 60 percent. A graph of AT&T's long-distance market share during these years closely resembles a ski slope.

The "Baby Bells" had no intention of walking away from the increasingly vicious battle between AT&T and MCI. Even before the 1982 agreement was formally signed, the Baby Bells were in Washington lobbying for entry into the lucrative long-distance market and in court arguing that AT&T management should not have been able to negotiate with the government on their behalf, and that therefore they should not be bound by the agreement. AT&T was soon battling for customers with the pieces of its former empire, and, as with every former imperial power, lingering resentments abounded.

For the newly established Baby Bells, AT&T was both supplier and competitor, an untenable position that gave rise to much hostility. The new long-distance carriers, MCI and Sprint (the latter's roots grew from a tiny Abilene, Kansas, telephone company established in 1899), had no intention of enriching their most reviled competitor and never bought so much as a single copper strand from AT&T Network Systems. Foreign equipment manufacturers, particularly Northern Telecom (later renamed Nortel Networks), used this period in the 1980s and 1990s to solidify their competitive position and grab market share. AT&T was in no position to retaliate effectively because, although the telecom industry spans the globe, the company had virtually no international presence. By the mid-1990s, market conditions were painfully inhospitable for Network Systems. CEO Allen was driven to examine the deteriorating situation.

Like his predecessors, Allen was initially loath to let Networks Systems go. Equipment manufacturing was very much part of the AT&T identity, part of its view of itself as a full-service provider, and part of its plans for life in the computer hardware business. For decades AT&T had fought the government's attempts to peel off Bell Labs and Western Electric. To voluntarily let them go would be like giving up territory once the battle was won. Bell Labs bestowed an intellectual glow that allowed AT&T to refute any claims of stalling innovation. Bell Labs gave customers the confidence, real or not, that AT&T was on the cutting edge of product development and that, while price might be a consideration for going to a competitor, technology was not. Top management continued to insist to both employees and the public that the synergies of vertical integration were a crucial element of corporate strategy. As the Baby Bells flocked to its competitors, AT&T watched its equipment business decline, aware that product innovations and fancy marketing would not make up for the fatal flaw caused by its vertically integrated structure.

McGinn had long hoped Allen would liberate Network Systems. Network Systems was the AT&T division that McGinn had run since 1993, and it sold and serviced switching, transmission, and fiber-optic products to telephone carriers (known in the business as network operators). If the equipment division remained embedded in AT&T, its business would continue to erode, almost regardless of his efforts, and his career might easily stall. Operating inside AT&T, Network Systems was faced with an intractable conflict of interest. Network Systems' customers, the local phone companies or the Baby Bells, were now AT&T's competitors, and the two sides were involved in a vicious advertising campaign to steal customers from each other. During this hostile period Carly Fiorina, president of North American operations in Network Systems, repeatedly confessed her frustration to colleagues. In any two-hour sales meeting with a Baby Bell customer she would have to listen to an hour and twenty minutes of abuse about AT&T, leaving a mere forty minutes to make her sales pitch. Along with McGinn, she made it clear to senior management that AT&T could no longer risk further erosions in Network Systems' sales to major industry players; in this context, the arguments for unshackling Network Systems became overwhelming. McGinn and his executive team had everything riding on Allen's decision and could only stand by and watch for years while Allen examined AT&T's options.

By 1994, Allen had decided to dispose of the Network Systems division. Twice during the previous four years he had called for formal reviews of the situation. After one review he asked the Management Executive Committee to take a vote on untethering Network Systems. Despite the fact that the business was operating at a suboptimal level, only three executives, AT&T president and COO Alex Mandl, then–Network Systems chief William Marx, and McGinn, voted in favor of severing the tie. Now Allen was ready to act. He asked Mandl, McGinn, and Fiorina to examine the idea of a merger between Network Systems and Swedish telecom giant Ericsson. For months the three studied the question and then gave Allen their unqualified backing. Allen and Mandl traveled to Sweden and presented the case for a merger to Lars Ramqvist, CEO of Ericsson. Allen told Ramqvist that a merger between the two would create a perfect fit. All of Network Systems' business was within the United States, where Ericsson's share of that market was insignificant. Ericsson was a major force in the wireless business, and AT&T was still predominantly a wireline player. Ramqvist and Allen walked away from the meeting in agreement that a merger of equals could benefit both companies. But the Wallenberg family, who held a controlling interest in Ericsson, was not interested, and Allen was forced to seek an alternative solution.

Although Network Systems continued to lose market share to its competitors, Northern Telecom, Ericsson, and Alcatel, there seemed to be little urgency in the search for a resolution. In the summer of 1995, Allen convened a number of meetings with his top management seeking their views on a possible spin-off of the network equipment businesses. Many of the executives were noncommittal, a few stated their position in favor, but the majority, until Allen's September announcement that he was doing it, were opposed to severing the tie.

Two exogenous events conspired to tip the balance. AT&T's computer business was losing almost half a billion dollars a year, cutting total corporate profits in half. Investors had weighed in with their verdict, and it had not been positive. Shares of AT&T were underperforming, and few analysts had anything promising to say about the future. "The market value of AT&T was being buried," Allen argued when he announced the spin-off. "Investors couldn't understand the strategy of the combined company." It seems more likely that investors fully understood that AT&T's competing businesses were harming each other and that, by moving into the computer business, AT&T had ventured far afield into a business in which it would never become a powerful force.

The bill that would become the Telecommunications Act of 1996, then being debated by Congress, was the sword of Damocles hanging over AT&T. Although the form and timing of the legislative upheaval were still uncertain in the summer of 1995, it was already clear that the impending transformation would alter the face of the industry in which AT&T now struggled. Similar legislative changes would be enacted in Europe, blasting open markets for American competitors to enter. Technological innovation was driving the agenda. The long-standing boundaries between long-distance and local calling were immaterial as the number of possible ways to communicate increased and the number of industry players escalated. When telephone calls could be made through computers or on cell phones, the role of the plain vanilla long-distance operator would shrink fast. The Telecom Act blew apart the industry and allowed poorly capitalized upstart competitors to enter the previously controlled local calling market. For Lucent, these legislative actions would mean everything; its business was profoundly and permanently altered. PR head Kathy Fitzgerald explained, "Consider that until 1984 we dealt with 'one' customer—AT&T. That became eight customers [AT&T and the Baby Bells] after divesture in 1984, and then fewer than a hundred service provider customers when we spun-off from AT&T. Today we have a thousand and added 650 service provider customers in the first half of 2000 alone."

AT&T was forced to change because of its own failures. From its cloistered seat in Manhattan, the company operated on what even its chairman called "Bell time." According to Allen, who would later describe himself as a "reformed monopolist," AT&T had taken seventy-five years to contemplate the introduction of colored phones. When Continental Airlines asked for an advanced phone reservation system in 1972, AT&T informed the company it would take eight years to build. The airline turned to Collins Radio, now part of Rockwell, which completed the project in two years. "And, while the rest of the world was experimenting with just-in-time manufacturing," Allen remembers, "we were still mired in a just-in-case mentality, designing phones built to last fifty years." AT&T had rarely acted as a force for change, and, despite its ownership of one of the world's premier research facilities, had a long and inglorious history of making changes only when they were long overdue. Changes came to AT&T from the outside, from governmental or competitive demands. For AT&T's management the goal was to proceed without disruption, to determine what it believed was best for the customer, country, and economy, and then to give it to them. AT&T would bring out a technological change only when it believed that the dislocation was minimal and the technology perfect, indifferent to the concerns of customers or competition. As comedienne Lily Tomlin's telephone switchboard operator, Ernestine, immortalized in her classic sketches, "You see, this phone system consists of a multibillion-dollar matrix of space-age technology that is so sophisticated even we can't handle it. But that's your problem, isn't it? Next time you complain about your phone service, why don't you try using two Dixie cups with a string? We don't care. We don't have to. [Ernestine snorts] We're the Phone Company."

For more than a decade the decision to separate AT&T's long-distance, computer, and equipment manufacturing businesses had hung in the balance, agonized over from every angle. In September 1995, Allen decided to slice AT&T into three unequal pieces, liberating the computer and equipment divisions to trade as separate public companies while AT&T held on to the profitable long-distance businesses. For the second time since 1982, the management of AT&T would dismantle the largest corporate edifice ever constructed, one that had taken more than a century to erect. Allen's decision was hailed as a bold move, one that would help AT&T and the businesses it was setting free regain the edge that had been eroding for decades.

AT&T was a textbook argument for a spin-off. The stock market had discounted its value as a conglomeration of businesses, a situation that began to reverse itself within minutes of Allen's announcement that he would spin off the two divisions. Within hours of the announcement AT&T's share price had risen 11 percent, and in its euphoria the financial markets added $10 billion to the value of the corporation. Schacht thought it was a brilliant decision, one that was not at all obvious at the time and for which he believed Allen deserved a great deal of credit.

In retrospect, there can be little doubt that AT&T waited far too long to make the inevitable change. Allen argued that AT&T operated from a position of strength in undertaking the trivestiture. "I'm trying to shape the future rather than react to it. This *was* the time. If anything, we're now ahead of the rules we play by," he asserted in his announcement. He hoped this restructuring would be viewed as a bold maneuver that evinced the company's foresight. After all, he had an image to set right as AT&T had a reputation as a plodding company. Perhaps action might prove him an innovative and courageous leader. But this was change AT&T style: wait until the problem is obvious to all, the stock market has rendered its judgment, and even some of your own management demands change, then try to make it look like a novel idea.

Many of Lucent's problems can be traced back to this delay. Lucent's management is still loath to criticize AT&T, as it remains among Lucent's largest customers. However, there were no more profitable years in the telecommunications market than 1996–2000, and because of Allen's lengthy decision-making process Lucent spent some of this time finding its feet. CEO McGinn was the first to admit that the postponement was a handicap. He later recalled, "We had to form a company, get a management team in place, get a name, get a set of books, a logo and do the IPO in a short period. So, we didn't spend the time to focus on a strategy for the business and we lost a year in getting into areas like data networking and software. Lucent certainly didn't start as a 98-pound weakling, but we were not ready to go full speed ahead."

For the employees of Network Systems, the spin-off generated questions. What would the new company be named? Where would it be located? Which businesses would depart, and which would remain with AT&T? Notably missing from Allen's announcement of the new company was any indication of who would run the venture. One reason AT&T's management delayed finding a solution for Network Systems was its concern that the division did not have substantial and seasoned enough managerial talent to operate as a separate company. As a former executive explained, "It was hard to see Network Systems as more than Western Electric grown up." Although later Allen would privately admit that he had underestimated the leadership talent that resided within Network Systems, in 1995 he thought they needed outside help. McGinn was appointed to lead the transition, but Allen left the door wide open to speculation when he said that he would announce his decisions on the new company's leadership at a later time.

CHAPTER THREE McGINN AND SCHACHT

Either you found Rich energizing or exhausting
—either stimulating or frightening.

Former Lucent senior executive

Every employee who came to Lucent by way of AT&T likes to think of himself as an outsider, a dissenter within the painfully conformist culture. Interview twenty Lucent/AT&T employees, and nineteen of them will explain why they are "not really AT&T types." The twentieth assumes that it goes without saying that anyone who was really an AT&T type would never have joined Lucent. But with Rich McGinn this distinction may in fact be accurate.

From McGinn's very first days, he was explicit in his intentions to create a wholly different entity: a high-tech growth company in an industry historically plagued by slow growth and gradually evolving product lines. To do this, Lucent would need to diversify away from its historical base of selling voice transmission products to telephone companies. While the voice business paid the bills, it was the data business that would be the source of innovation and growth for the industry. We are going to be a data networking company in a very short period of time, McGinn never tired of telling his troops. An intellectual and a strategist, McGinn had plans for Lucent that would, if he succeeded, leave it unrecognizable as an AT&T offspring.

Despite three decades with AT&T/Lucent, McGinn was not *of* AT&T.

In a disparaging comment that, in fact, highlighted what may have been McGinn's greatest strength, a senior Lucent colleague explained, "Rich tried to turn Lucent into the company he wanted to be chairman of—not the company it was." McGinn was not the obvious candidate to remake Lucent; there was nothing in his résumé to signal the changes to come. His father was a commercial artist for the New York *Daily News* and his mother worked as a secretary at AT&T. Neither a scientist nor an engineer, not even a student of business, the sports-mad McGinn graduated from Grinnell College, a Congregationalist college in Iowa, in 1969 with a history major and the vague notion of attending business school at some later date.

Fiercely competitive, McGinn claims, "I never met a sport I didn't like." Youthful and athletic, he seemed to thrive on competition. A fish story he relishes telling is revealing. On the second day of a three-day fishing trip, McGinn hooked a 120-pound tarpon that while still on the line managed to entangle itself with a dock pylon. Rather than lose the fish, McGinn dove off the boat into the water and swam around the pylon, still clinging to the rod. The fish dragged him out to sea and his boat caught up with him and pulled him back on board. Witnesses to this spectacle thought he was mad, leaping into shark-ridden waters, his hands already sliced up from the battle, but "I got the fish," McGinn recalls.

After college, McGinn headed for the management trainee program at Illinois Bell and scaled the management hierarchy of the regional carrier. He jumped over to the mother ship and in 1978 joined AT&T's corporate marketing organization, where he proved facile at climbing AT&T's rigid corporate ladder. In 1982 McGinn moved to AT&T International, and by 1987 he was in AT&T's Data Systems Group with responsibility for strategy and business development. (He was later named president of the group.) In this capacity, McGinn was the driving force behind AT&T's farsighted investment in Sun Microsystems. On the same day in 1987 that the stock market crashed 500 points, AT&T was signing an agreement to buy 20 percent of Sun Microsystems. McGinn hoped it would catapult AT&T into the computer business. But it was an investment on which AT&T failed to capitalize. AT&T's computer business was not a success before, during, or after McGinn's leadership, and in 1991 he transferred to the telephone equipment side of the business.

By 1995, the forty-nine-year-old McGinn was CEO of Network Systems, lived one town away from where he had been born in New Jersey, and, although a married man with children, had dinner with his mother every Sunday night. He was an unusual mixture of a deeply loyal company man and a risk taker with ambitions to challenge the tyranny of the status quo. McGinn's loyalty ran deep. In his years at Lucent and afterward, he never sold a single share of Lucent stock. At a time when many CEOs were cashing out on their bounty, McGinn remained fully invested in Lucent. Over the years, McGinn continually exhorted his board to buy more stock and pleaded with them to hold on to the shares they held. He even invested his elderly mother in Lucent stock, diversifying her portfolio with a bit of Dell Computer. David Nadler, a consultant with very close ties to Lucent for many years, explains, "He was always part of the outside renegade culture at AT&T."

The Network Systems sales force, at least those who were most successful in their peddling, had already witnessed firsthand the daredevil side of McGinn. At AT&T, and later at Lucent, the company's top tier of successful salespeople—those who had reached 100 percent of their annual sales quotas—were invited, along with a guest, to the world's greatest resorts—gratis. As members of the Achievers Club, Super Achievers Club, or most-exclusive Leadership Council (the top 2 percent), they would be entertained at business conferences in which the company's leaders would address them, but the focus was on fun. At locales from Hawaii to Mexico to Nevis and Aspen, these sales people were bathed in luxury. Some of those who attended these celebrations of success remember them as among the great events in their lives. The days were full of sports and spas. The hotels were first class and the meals were lavish. In the evening they were treated to black-tie dinners with famous entertainers, but perhaps more important for those who had shown the company years of loyal service, these were opportunities to break bread with their leaders. After dinner there was always an awards ceremony in which each salesperson came onstage to greet top management and receive recognition for his or her bumper year. "Being brought up in front of your peers like this, for salespeople, was as good as the cash," recalls one sales manager.

Nothing was more entertaining than watching McGinn publicly place himself in daring positions, with perhaps a bit of danger involved. At each conference he made an even more spectacular entrance, once roaring onto the stage on a motorcycle. Everyone knew that although McGinn was a self-proclaimed adrenaline junkie, he was moving "way out of his comfort zone" to please the crowds. His willingness to poke fun at himself and his management team endeared him. At one Achievers Club the theme was racing, and McGinn joined the professional drivers and his employees racing around a track at speeds that at least gave the impression of very real danger. Later the videotapes would be replayed for the entire gathering in case anyone had missed the race. On another occasion a video was shown of McGinn in the role of a superhero, dressed in a tuxedo and cape, who flies across the globe solving communications problems with the help of his trusty management team. As the show ended and the lights went on, McGinn stood ready to address his audience wearing the very same costume.

McGinn stood in awe of the prospect placed before him in 1995. He knew that Lucent was his great opportunity and was determined to lead the company through a cultural revolution. "I was and remain surprised by how quickly our people have adapted to the new world," he said shortly after taking charge. "What we need here are the kind of people who enjoy coming in in the morning, kicking the door down, and going after the excitement out there in the market, the new technologies. How many times do you get a chance to take a $19 billion company and make of it what you want it to be?" McGinn knew that a clean page would be his only once, and with his assembled team he tried to create a shared vision of what they would write. "Rich gets people to have expectations of themselves that are greater than what they would have on their own," said Schacht. But McGinn recognized that he fought a difficult battle. The AT&T culture was deeply embedded in his employees. He and some of his management team enjoyed the thud of a door being flattened against the floor, but they wondered how many employees were frightened by this sound.

Those who were interviewed by McGinn for senior positions have indelible memories of these meetings. McGinn would often arrive, sometimes straight from the golf course, dressed in casual clothing, usually no tie, often no socks. McGinn disparaged more formal dress and jokingly insisted that ties cut the flow of blood to the brain. Committed to bringing in high-level talent from outside, he would block hours out of his schedule to spend with a prospect and with incredible eloquence outline his views of the industry and plans for his company. McGinn's conversation jumps around easily; he quotes Churchill one minute and research scientists at Bell Labs the next. His deep, resonant voice betrays no signs of a regional accent, and with his dark hair and usual black shirts he looks more like a college professor than a corporate CEO.

Lance Boxer worked at MCI for seventeen years, and if there was one lesson he learned during his tenure, it was "never do business with AT&T." Even after Lucent was a stand-alone company and Fiorina was assigned to cover MCI, little had changed. So when she suggested to a headhunter that Lucent would be interested in speaking to Boxer about a position, he was reluctant to consent. Boxer, the fortyish chief technology officer for MCI, thought of AT&T and Lucent as a large, staid organization that he was not interested in as either a supplier or an employer. Yet after a two-hour conversation with Dan Stanzione, head of Bell Labs, Boxer was softening. "Stanzione was one of the smartest men I ever met," Boxer recalls, "a man of huge integrity, a real steady Eddy at the company." Then Lucent went silent. For two and a half months Boxer heard nothing, a period that just confirmed his worst fears about the organization. He knew that when MCI wanted an executive, it hired him. It might take a week, ten days, but the process moved steadily forward. By the time the headhunter called back and said, "They want to make you an offer," Boxer was fed up and told him, "I'm no longer interested, and I certainly do not want to work someplace that takes two and a half months to make up their mind." Although Boxer thought he had heard the last of Lucent, the headhunter quickly returned with the message, "Rich McGinn wants to see you. He doesn't believe 'no' should be an answer."

As Boxer boarded the corporate jet that McGinn had sent to McKinney, Texas, he was uncertain about what to expect. McGinn was a superstar in the industry, with a flawless reputation. As Boxer approached Lucent's monumental lobby, he was awed. "Chills ran down my back when I thought that the very essence of how we communicate came from this building," he recalls. Boxer was shown to McGinn's third-floor office, and when the CEO entered the room he simply put out his hand and said, "Hi, I'm Rich. Let's talk about Lucent and why you should be here." McGinn is a man comfortable with himself. Seated with his feet up on his desk, chair tipped back precariously, he exudes an easy confidence.

Hours later, Boxer emerged from the meeting convinced. McGinn had spoken to him like an old friend, Boxer remembers: "He was brilliant, articulate, funny, but a real guy. He was also a man on a mission, to hire the best, grow the company, and take on the industry. Rich was the kind of guy who decided something and then stopped at nothing to get it. He was very focused, and that was part of the allure. I thought Rich was exactly the kind of leader I wanted to work with." Boxer's formal offering letter was in his hand forty-eight hours after he left McGinn's office, and he quickly accepted. The whole process had changed Boxer's view of AT&T. He says, "If you wanted to work in this industry, Rich made you believe that Lucent was the place to do it, almost like there was magic in the air."

The AT&T spin-off was a little like a divorce, complete with some of the nastiness and recriminations that come with the territory. The settlement was an especially heated process. Negotiating issues of intellectual property and shifting the corporate debt burden proved to be extremely sticky. AT&T's chief concerns were for its own financial health, not that of its soon-to-be-former employees. McGinn argued Lucent's case, as he saw that AT&T was strengthening its balance sheet on the new company's back. Yet each time he pushed for what he believed to be fair, he was told, no, this is not a negotiation. In the end Lucent was loaded with debt, stripped of its accounts receivable, and burdened with extraneous businesses like the phone stores. Those who left for Lucent remember that their AT&T colleagues thought they were crazy to join the insurgents. AT&T was a proven entity, and the untested Lucent might go down in flames. To justify secession from the mother ship and the trust of those who went with him, McGinn would need to make Lucent a triumph. Mere success would not do.

Late in the afternoon of September 19, 1995, McGinn telephoned his top executives and invited them to a special meeting in the conference room. In a gesture that even some of his detractors still find touching,

McGinn greeted them with beer, pizza, and a huge smile. He was the bearer of good news. He looked around the room and told them they would have the chance to run their own company; they would be the senior management team and the challenge would be theirs. They needed to understand that this would be a complete break with the past.

That night and over the following three weeks, McGinn's team met to take stock. He told them that they would need to move at breakneck speed after years of deliberation and that they needed to "get yourself prepared for a world where expectations for us will be much higher." Together they examined each of their businesses, looked at its prospects, and outlined budgets, priorities, and staffing plans. It was an unsettled time. McGinn had not been named CEO, although it was assumed by all that he would be. The new management team did not know exactly which businesses would be theirs to run or which resources and staff would make the move. But it was also an extremely exciting time for a group of executives who had long been buried under layers of AT&T bureaucracy, unknown outside of New Jersey. They were about to become national news.

McGinn was determined to carve a high-tech powerhouse out of the vestiges of a regulated monopoly, something for which there is little precedent. With the pizza still hot and the beer still cold, he reviewed the financial feats of Hewlett-Packard, Cisco, and Motorola, and posed the questions that would determine much of Lucent's direction over the next six years: "Why can't we too grow at the same pace as these companies?" and "Why can't we earn from the market a [price/earnings] multiple like they have?" They were audacious questions from the unanointed leader of a not-yet-independent company that had lost money four out of the previous five years. But still McGinn insisted, why not?

On September 20, the day of Allen's announcement, McGinn met with Network Systems' employees in the cafeteria. He entered the room to ringing applause and shouts of praise. Charismatic and popular, he was a leader with whom the new Lucent was already comfortable. Most of McGinn's troops believed that because he was the departmental CEO, he would continue as their leader once the spin-off was completed. McGinn, too, was confident that Allen would hand him the company, just as he had placed him in charge of the transition. He had considered the possibility that Allen would name a nonexecutive chairman to oversee the transition to a public company, but McGinn felt certain that it would be his show to run. For three weeks, the press, which had been left hanging on Allen's phrase that he would name the top management "at the appropriate time," grilled McGinn on whether he had been secretly anointed. In October, *Time* magazine erroneously referred to McGinn as the chief executive of the as-yet-unnamed equipment manufacturer. In this exceedingly awkward situation, McGinn could do nothing but dodge press questions or give carefully worded answers that amounted to saying nothing. McGinn knew he was up to the task; he was just waiting for the nod.

But Allen had doubts. McGinn's relative inexperience made him uneasy. He did not have confidence in McGinn's ability to lead a public company on his own and presented his concerns to the AT&T board of directors. Bill Marx, who had previously been McGinn's boss, was one of Allen's regular golf partners. Marx, who close colleagues say tried to land the top job himself, expressed his doubts to Allen about McGinn's relative inexperience. For the new company to have sufficient credibility with investors, analysts, and the enormous shareholding public that would be created by the spin-off, Allen believed that it needed to be led by someone with substantial experience at the helm of a publicly traded company. McGinn had shown himself capable of running a division, but a separate public company, Allen felt strongly, was an entirely different matter. McGinn might know the Network Systems business, but in Allen's view he would not lend credibility to the new entity. Allen considered bringing in an outsider, someone with CEO and industry experience, but decided he needed someone from closer to home.

In the midst of the board's deliberations over the new company's leadership, Henry Schacht, former CEO of Cummins Engine and for fourteen years an AT&T board member, was called out of the meeting to the telephone. Schacht was away for about forty-five minutes, during which time Allen spoke at length on the leadership qualities he felt the new company needed. Allen was looking for an older statesman, preferably someone in his early sixties, who could act as a mentor for McGinn. Legend has it that as Allen enumerated the qualities he was looking for in the new CEO, including prior CEO experience with a manufacturing company with an international base, one of the board members turned to him and said, "We have someone like that here." Some close to Allen have suggested that the meeting was staged; Allen had wanted Schacht from the start and gave his list of qualities in order to point out to his fellow board members the many reasons Schacht was the ideal candidate.

Henry Schacht had long been considered an éminence grise of the corporate world, yet during the autumn of 1995 he was winding down his career. For two decades he had run Cummins Engine, the leading manufacturer of diesel engines. Slim, tall, gray-haired, and always dressed conservatively (McGinn once joked that Schacht was one of Lucent's founding fathers and looked the part), Schacht had retired at the age of sixty. Only months earlier he had left Cummins, his reputation at its apex. Schacht's leadership pedigree was perfect. At age thirty-four he became the president of Cummins and at forty-two the chairman, serving eighteen years before retiring. Yet it was more than just leadership that Schacht had exhibited. He was the establishment, and for Allen the low-risk option. Both men were sixty, transplanted midwesterners who sat on the boards of an array of major U.S. corporations. Schacht had graduated from Yale and then gone on to Harvard Business School; McGinn had only his undergraduate degree from Grinnell. Schacht held positions on the boards of CBS, Chase Manhattan, and Alcoa and was a trustee of Yale, the Metropolitan Museum of Art, and the Ford Foundation, which made him a member of the business elite, a man on a first-name basis with hundreds of CEOs. McGinn's affiliations were largely confined to AT&T.

Yet appearances deceive. The unassuming Schacht, while looking every bit the part of a conservative patrician, had had a very different upbringing. When Schacht was in the second grade, his father, a fisherman, walked out on his wife and child. Raised alone by his mother, a buyer at a small store in Erie, Pennsylvania, Schacht sold eggs from hens he raised to help with the family finances. Deeply thoughtful, Schacht lays out his arguments from first principles, every argument carefully reasoned and presented. He builds his arguments slowly, never jumping from A to C and assuming his listener will fill in the blanks. After an extensive discussion of the characteristics essential to the CEO of the new corporation, the AT&T board members agreed that Schacht was the right man for the job. When he returned to the board meeting, he asked how the discussion had progressed, and his fellow members laughed.

The cautious Schacht asked Allen for a week to think about the board's offer before accepting it. For both Schacht and McGinn, this was an unparalleled opportunity. Schacht had retired relatively young from a company only a fifth of the size of Lucent; this was a chance to star in an entirely different league. "For Henry this was a huge deal," recalls a former AT&T colleague, who regarded Cummins as a mixed success. "He probably pinched himself every morning and asked, 'How did I get here?'" Schacht's efforts to face down Japanese competition had ultimately prevailed, but Cummins was a much smaller company that had endured many years of losses along the way. For McGinn, who had presided over AT&T's failed computer business, then labored under less-than-ideal conditions at Network Systems, Lucent offered the chance to show he could run a highly successful company.

Schacht's new position was not a total shock to McGinn. At a recent board retreat in Greenbrier, West Virginia, Allen had thrust McGinn and Schacht together, in meetings and at meals, and McGinn suspected that something was up. At 10 A.M. on October 12, 1995, the taciturn Allen sat McGinn down and offered him the COO and president's titles. Allen said, "I would like you to work with Henry as a team. Why don't you go talk to him." McGinn protested and told Allen that he was up to the CEO job. Allen replied that the announcement was scheduled for one o'clock; did he want the COO position or not? In his press release, Allen hailed Schacht's leadership as well as his "experience—especially in the application of technology." He may have taken it too far. Schacht was drafted because he knew how to run a company, not how to build a network. Schacht's mission was to establish the spin-off and cement the new management's credibility. He would be a guiding figure to a company full of fearful and uncertain employees. Perhaps the most important part of his job went unstated; as he later confided to consultant David Nadler, "I see a big part of my job here as helping to make Rich the best CEO he can be someday." McGinn was the CEO-in-training from the start. Whenever he and Schacht could not be found, employees joked, "Rich will be back later, he's at CEO school." But McGinn was not on trial. Schacht made this clear from the start. He recalls, "We started off saying, 'Rich, there is no alternative in the company to you to be the next CEO. My job is to do everything I can to make sure you are the next CEO. We're not going to hire somebody else to make this a horse race.' " Schacht told everyone, "Look, Rich is my guy. I'm here to help him lead the group."

"This isn't what I had in mind," McGinn informed his new boss at their first face-to-face meeting. "I didn't think it was," Schacht graciously acknowledged. Although McGinn had been severely disappointed in Allen's decision, by the time he met with Schacht in a small office at AT&T's Basking Ridge, New Jersey, headquarters he was looking for a way to make the arrangement work. Allen had given McGinn's business to a man who knew little about telecommunications, knew only a handful of Network Systems' employees, and had overseen a $4.2 billion company, no bigger than a department inside a division of the soon-to-be $20 billion Lucent. One senior manager remembers thinking, "How is Rich going to pull this off with an old guy in a gray suit who no one in high tech has ever heard of?" Schacht offered a proposition: theirs would be a partnership. He promised that he would work to make McGinn feel as though they were operating as equals. He told McGinn that he hoped they could talk through this stressful situation.

Each man needed the other. Schacht acknowledged that he was the CEO of a company in which he knew little about the marketplace, technology, industry, management, and competitors. He would need to rely on McGinn's experience and knowledge of its customers and employees. "At first we couldn't figure it out," remembers Marc Schweig, a senior vice president in sales, echoing the sentiments of many at Lucent. "Of all people, why Henry? Why launch a high-tech company with a semiretired guy that ran Cummins Engine and had no visibility in the high-tech world? This was a guy who admitted he knew nothing about the industry. But within days it became obvious that his raw business sense, leadership, and integrity were essential." Others found Schacht comforting, a CEO straight out of central casting, a man who looked and sounded the part and had the résumé to back it up. McGinn still retained the option to walk, but after twenty-six years with AT&T, he was close enough to taste the prize. At their initial meeting he told Schacht, "Henry, I propose this. We have six months in which to go public and an incredible amount of work to be done. Let's talk about this again one year from today." Schacht protested and said they should address their concerns now; this was a chance to clear the air, to start off on the right foot. But McGinn insisted, and it was agreed. They would resume their discussion of Lucent's leadership one year hence.

From the start, Schacht very publicly referred to McGinn as "my partner." "Henry-Rich, without the verbal ampersand" was McGinn's description. Later, when McGinn recalled the situation, he praised Schacht: "Henry reached out to me. He said we could work together, that he didn't believe in a division of labor, and he had proved that for more than twenty-five years at Cummins." Schacht's conviction extended to both the substantive and the symbolic. He refused to allow his picture to be taken for any publicity without including McGinn in the photo. McGinn flew into Manhattan by helicopter for an interview with *The New York Times* because Schacht refused to let the story run unless McGinn's face appeared on the paper's pages. Dozens of their current and former employees attest to their partnership and confirm that they ran the company together. From the beginning it was understood internally that Schacht's tenure would be brief. As he acknowledged, "We were into management succession the day I walked in here." To the outside world theirs looked like a smooth transition of power, but it was made possible by a series of behind-the-scenes deals.

Three weeks after McGinn had given his team the good news, Schacht stepped into the CEO role and set an entirely different tone. Instead of immersing themselves in the business of running the business, Schacht asked his executives to step back, contemplate the array of opportunities a blank piece of paper provided, and think long and hard about the mission and expectations for the new, as-yet-unnamed entity. In his view, an opportunity to establish the company's values and beliefs came along only once. "My advice to anybody who wants to be a change agent is to do the reverse of what everybody says," Schacht explained. "Everybody says, 'Get in there and get busy.' My answer is: stop. Figure out the lay of the land, what it is that needs to be done, and how you're going to be the leader."

Within weeks Schacht's team, AT&T executives from all of the divisions that would make up Lucent, had grown alarmed at what they perceived to be his lack of urgency, the step-by-step approach to meetings that often sounded like something straight out of a business self-help book. Schacht could be pedantic, often lecturing his fellow executives about his experiences at Cummins Engine. And while they certainly had faith in his leadership abilities, there were serious reservations about the relevance of the diesel engine business. Time was a-wasting. Surely, Nortel's and Cisco's managements were not sitting around wondering who they wanted to be. This was a luxury Lucent's top managers believed that they could not afford.

Schacht knew of their impatience but felt certain they were doing something they could not afford *not* to do. He had a discordant team of managers who shared nothing but their AT&T past; in his view, without some drastic changes in their outlook they would not be able to successfully run a corporation together. He believed that a major part of his role at Lucent was to be a teacher. This meant that his greatest contribution might not be any particular action but rather imparting management lessons that would outlast him.

More than a corporate structure, marketing plan, or departmental budgets, Schacht believed his new company needed a basic statement of mission and values that would be drafted and adopted at the highest levels and then disseminated throughout the organization. Without this bedrock, Lucent could too easily go adrift. "The value system of the corporation will become the primary, if not sole, anchor in the destabilized world we are entering." Schacht said.

Meeting with the top sixteen managers over a matter of months between October 1995 and April 1996, Schacht guided the development of these documents. Yet his top management was his most skeptical audience. In one of the early meetings, Schacht asked each executive to name a word believed to represent the new company; then they went around the room and named another and another. Finally the group looked at the words most often repeated and discussed their relevance. From this the mission statement was drafted, with the entire group arguing about the inclusion of each word. It was a lengthy process by which every person came to understand every other person's perceptions of the new organization. Schacht viewed this as an essential team-building exercise. Unless his management knew what they stood for, how would they know where they were going? His team did not necessarily see it this way.

In fact, from the start Schacht's troops were in revolt. They brought their cell phones into meetings and arranged for their secretaries to interrupt, finding any means possible to get back to "work." Daunted by the operational tasks that lay before them, no one wanted to spend his days sitting in a conference room voting on the language that would soon be buried in the annual report. So Schacht took his team off site and locked the doors. He knew this would get their attention. As he recalls, "After six or seven half-day meetings, one of the most skeptical older hands stood up and said, 'You know, why don't we just finally admit it; what we're all just suddenly realizing is that this is an opportunity of a lifetime and we shouldn't mess it up.' " Schacht had repeatedly emphasized that the new company needed a sense of urgency, momentum, and speed, but many on his team felt these protracted sessions suggested the opposite.

Schacht wanted his team to form new identities. Managers and their employees would no longer be "Western Electric" or "Bell Labs"; they would be the new company. As heads of formerly disparate departments within AT&T, the assembled managers had operated with a "stovepipe mentality" (feeding profits and information upward), far more concerned about the results of their divisions than about the company as a whole. Now connections would cross divisions and senior managers would need to concern themselves with performance outside their own departments. Executive compensation would be biased toward the results of the entire company rather than any one division, but, even so, Schacht believed that unless a set of shared values and benchmarks could be agreed upon, cohesiveness would remain illusory. In one exercise, after the company's name had been chosen, Schacht asked executives to don a hat labeled with the names of their divisions. Then they all left the room, to return immediately and exchange their old hats for ones labeled "Lucent." Many on his team viewed this as an ineffective use of their time.

Schacht recalls, "We spent an average of one day a week over the first six months pounding these [values and mission statement] out in closed sessions plus countless hours testing the relevancy and validity with groups throughout the new corporation, making numerous corrections and refinements as a result of the ownership buildup process." Some of those who sat through these meetings remember a great deal more time spent this way. A number of top managers disagreed with Schacht's approach. They believed that AT&T had been too preoccupied by what happened internally. One senior executive was very worried, observing, "This was not the right focus. Lucent needed to be more focused on customers and the outside world—but a long series of internal meetings were not helping this."

Schacht probed some of the failures of the past in these early meetings. Network Systems had not been a roaring success, and the less-than-profitable areas left him wondering. "Why didn't you invest in a data company to support your Business Communications Systems business?" he asked. "Why did you keep certain factories open when the business wasn't growing or profitable?" The frustration was palpable, and the response was always the same: "They wouldn't let us." Schacht listened closely as his team reeled off a list of roadblocks that had been placed in their way by AT&T, and then he paused, walked over to the whiteboard, and wrote two phrases that seven years later still linger in the memories of everyone who was in that room. "THIS IS DIFFERENT," he wrote. "WE ARE THEY." In a moment that resonated with everyone who read his words, Schacht put the past in its place and changed the tone of everything that would follow. AT&T would no longer be a punching bag; Lucent was on its own.

Although senior managers still remember this time with some frustration, they acknowledge that the team-building exercise was extremely valuable. As one explained, Lucent is a far better institution for having clarified its values and goals, although the "sustained amount of time spent was unrealistic." These early management sessions achieved Schacht's three objectives. First, a cohesive, highly motivated team grew out of the disparate managers who had been tossed together by their previous employer. Second, a sense of identity emerged from a shared vision, and a sense of excitement replaced any trepidation about leaving AT&T. Finally, and equally important, the level of expectations was permanently raised and no illusions remained about slipping back.

"Everything is being rethought, from how we deal with snow days to our plan of attack in new markets," said Network Systems vice president Carly Fiorina. "It's been exhilarating and exhausting." In this new world, AT&T's old standard of performance, simply to improve upon the previous year's results, would be woefully inadequate. Schacht and McGinn set a new, altogether more challenging goal, to be "best in class." They intended Lucent to exceed the standards for product innovation and quality, financial success, and customer service set by its competitors in the telecommunications industry. Each of the sixteen senior executives was instructed to go out and study the best company in his or her respective area, as well as other highly successful companies in similar industries. (In what would prove to be ironic, Fiorina was assigned to study Hewlett-Packard—the company she would later leave Lucent to run.) Going forward, they would benchmark not against their own historical results but against those of the icons in their respective industries. For many Lucent executives this was an eye-opening experience as they examined cost structures, productivity (revenue per employee), and profitability that put Lucent and AT&T to shame. Cloistered inside AT&T with their customers firmly tethered to their sides, Network Systems executives had done little to survey the industry landscape. Schacht and McGinn needed to alter the sights of their team. Lucent's ultimate goal now was nothing less than to outperform the best of what they saw. Although this does not sound like a radical practice, for a company that for a century held itself above competition it was a truly revolutionary thought.

Not everything at Lucent was new. Despite the almost unending rhetoric about a fresh beginning, the newborn company would be led almost entirely by seasoned AT&T executives. In 1995, 80 percent of Lucent's employees had never worked anywhere but AT&T, and for most their experience did not extend outside the narrow confines of their small corner of the telephone company. All movement at AT&T was vertical; few ventured outside their own organizational stovepipe. AT&T was uncomfortable with outsiders. It was accepted as a truism that no one from the outside could understand the telephone business. Lucent was born with this bias. Newcomers described a distinctly insular culture where length of service was revered. AT&T had always been an up-from-the-bottom culture, and at first Lucent seemed essentially the same.

Schacht did not believe that Lucent needed to bring new managers into its top ranks. He thought that those who had shown loyalty should have a chance to run the new company, and he had faith that they would prove themselves. From his observations as an AT&T director, Schacht believed that Lucent possessed the necessary talent. Twelve of the top managers on his team in 1995 had been with AT&T for five years or more, and, like their employees, most had spent decades with Ma Bell. McGinn's views on this, although largely unstated during their joint tenure, differed quite markedly from Schacht's. McGinn set as his goal that within two years, 40 percent of the organization would come from outside, through either individual hires or acquisitions. When difficulties arose, McGinn would look for outside talent to bring a new perspective to Lucent's troubles. When Lucent had to report weak financial numbers to the market, he hired a new CFO. Later, when the optical business foundered, he placed the CEO of a start-up at the helm of this pivotal business.

Over the years Schacht relied on the same small group of officers, rearranging their roles as Lucent's fortunes shifted. As late as 2003, after Lucent had been engulfed in more difficulties than any company that had not filed for bankruptcy, Schacht never deviated from this viewpoint. Even after its seven years as an independent company, only one of the top eleven executives had never collected an AT&T paycheck. Schacht held steadfast in his belief that those who know the business best, who are committed to the organization and its plans, are in the best position to lead a company through periods of uncertainty. As an AT&T director, he felt that Network Systems had a store of underutilized talent that had been held back by a lack of opportunity. He points out, with understandable pride, that former Network Systems executives now run Hewlett-Packard, British Telecom, Avaya, and Agere, in addition to Lucent itself. But Lucent was entering a new world complete with new businesses, customers, and competitors. Competition would crawl out from under garage doors or splinter off of larger corporations. Lucent did not know this world, and Schacht was reluctant to recruit senior executives who had been there.

When the AT&T board sent Schacht over to the new company, he and Allen handpicked four AT&T board members to join him. In addition to himself, they would become the core of Lucent's new board of directors. The group included Carla Hills, former U.S. trade representative; Drew Lewis, chairman and CEO of Union Pacific; Franklin Thomas, former president of the Ford Foundation; and Donald Perkins, former chairman of the grocery chain Jewel Co. All were seasoned businesspeople with little experience that was in any way relevant to Lucent's businesses. These were the people who were going to have to oversee management with little background relevant to Lucent's operation. Every board member save McGinn was over sixty, and most were over sixty-five. Certainly their many years of business experience proved useful to management, and Franklin Thomas in particular would become an invaluable sounding board. When Schacht later expanded the board, he invited Paul Allaire, fifty-eight, CEO of Xerox; Paul O'Neill, sixty, CEO of Alcoa; and the most relevant appointment, John Young, sixty-four, former CEO of Hewlett-Packard. However, Lucent passed up the chance to bring to the highest levels of corporate governance venture capitalists or other financiers, researchers or academics, or others from the start-up world. Lucent's board did not resemble either its competitors' or those of the companies it sought to emulate, but rather a traditional Fortune 100 manufacturing company. Schacht felt the depth of executive experience on the board was an invaluable resource. But the board was constituted of his peers, men (and one woman) of his era, who saw the business in a similar way. Even later, after McGinn inherited this board, it would in many ways remain Schacht's.

CHAPTER FOUR 1996

Opening bell stock price: $27 (split adjusted $6.10)
1996 Stock price (split adjusted)   High: $10.17   Low: $5.70

> *Don't you mean "Lucid"?*
>
> Lucent employee upon hearing the company's
> new name

When Henry Schacht and Rich McGinn unveiled Lucent's new name in a televised internal broadcast to their employees, it was a historic moment. On a cold sunny day, February 5, 1996, a 127-year-old company was reborn. A nineteenth-century manufacturer would begin a new life as twenty-first-century start-up, a dichotomy that explains so much about Lucent.

Kathy Fitzgerald, senior vice president of public relations and advertising, moderated the two-hour show, cutting from snow-covered Murray Hill to sites in Texas and the Netherlands with the aplomb of a CNN anchorperson. At each location she garnered reactions and viewpoints, with the Texans shouting a huge "Howdy!" and those in Murray Hill sending back a "big New Jersey" welcome. Broadcast on the day that Lucent became a separate, if wholly owned, subsidiary of AT&T, the presentation was strangely reminiscent of a game show or awards ceremony, the conclusion of which actually involved opening a three-foot-long envelope wrapped in a red ribbon that contained the company's new name.

This was Schacht's first major public opportunity to underscore McGinn's role as his partner in front of their employees and the press. To do this, he stepped out of the limelight and had McGinn unveil the company's new name and logo. Even then, McGinn did not forget to credit his partner when he said, "The company would have the brains of Albert Einstein, the body of Hercules and the heart of Florence Nightingale . . . so we're naming it"—long pause, while McGinn grinned—"Schacht."

Determined to remake the new company's identity, Lucent's management had conducted an intensive search for a brand and an image that can only be described as the anti-AT&T. "When you bear one of the best known names in communications, selecting a new name takes on even more significance," Schacht explained. In April the IPO would take place and AT&T would sell 17.6 percent of Lucent to the public, in order to raise some cash but stay below the 80 percent threshold that would have to be given away to shareholders to ensure that the spin-off would be tax-free for them. Then, finally, in September, AT&T would spin off the remaining Lucent shares to all of its shareholders. The name search had been accelerated because the uncertainty created by the spin-off would only be exacerbated if 125,000 employees could not even name their employer. The naming process went forward in total secrecy because Securities and Exchange Commission guidelines forbid widespread communications about a new company prior to an initial public offering. A San Francisco brand-consulting agency, Landor Associates, was given twelve weeks to conduct interviews with management and focus groups for employees all over the globe and return with the brand. Landor and the Lucent and AT&T teams assigned to work with the agency agreed that they would need to spend seven days a week, including the Christmas holidays, if they were to develop an identity and launch it to the public. It would be one of Lucent's first tests of speed.

At the February ceremony, McGinn recapped for the company's employees the machinations management had endured to arrive at the final name. He explained that 11,665 companies have "net" in their names, 12,742 have "sys," 10,327 have "tech," and another 7,909 have "tel" (although he noted that most of them were not in the telecom business and wondered aloud what they were thinking), and announced that the new company's new name would be "Sysnettechtel." His audience loved it. McGinn told them that the employee focus groups had argued overwhelmingly against using abbreviations and this dashed the best names they had worked out for the three-way breakup, AT&T, BT&T, and CT&T.

The search for the new company's name created an inevitable controversy over what should be kept from the past, a debate that would recur with almost every aspect of Lucent's new business. There were two schools of thought. One favored a complete break with AT&T and the creation of an entirely new name—what the marketers liked to call an "empty vessel"—perhaps through the use of an invented word, a trend popular in the late 1990s. The other, which included McGinn, argued for a name that would evoke AT&T and the Bell System's stature, reliability, and integrity. He felt that the use of the Bell name would give the company greater recognition and instant legitimacy. McGinn led a contingent that favored calling the company AGB for Alexander Graham Bell, or possibly American Bell. But the AT&T people felt strongly that the Bell name should not be used. The 1984 consent decree, which had broken up the telephone monopoly, forbade AT&T the use of the Bell name, and many believed that the new company would be included in this prohibition (although lawyers scrambled to find an answer to this question as the process unfolded, this would remain an unknown). Another contingent believed that the name "Bell" evoked images of telephones that spoke to the past and a research lab that was slow to deliver products to market.

Landor sought out a name that that would communicate light, creativity, and connectedness. It had to be a name that worked all over the world and did not carry any sinister meanings in other languages. It wanted the name and logo to stand out from the new company's competition, but not so much that it looked strange. Through a painstaking process, Landor considered more than seven hundred possibilities and AT&T employees suggested a further four hundred. The name needed to shed some of the AT&T baggage, especially the slowness, inflexibility, and arrogance with which the company had been associated. At the same time, the new name needed to bask in AT&T's strong points of re-liability, stature, and experience. It also needed to conjure qualities the offspring company hoped to grow into: speed, energy, vision, flexibility, innovation, and a focus on the customer—all watchwords of the 1990s. The new name would need to convey that this was an entirely new com-pany with only some of the attributes of its progenitor. It was a lot to ask of a single word.

No one really loved the name Lucent. Schacht supported it and said that it suggested clarity of thought and had a technological feel. McGinn felt that "Lucent" had freshness and avoided the phony computer-generated sound of many corporate names. The choice of both name and symbol were controversial until the last, and Landor continued to de-velop the AGB name and an alternative, Telascent or Telescent. As the process dragged on, McGinn and Schacht chatted in front of an elevator one evening. They had reconciled themselves to the fact that the Bell name would not be available, that no new name would emerge to sweep them off their feet, and as their elevator car descended to the ground floor the two agreed that Lucent was the best option available. Yet neither was overly enthusiastic about the name, and as they ascended the stage to an-nounce it to their employees, McGinn turned to Schacht and said, "Come on, Henry, one last chance, we can still change it if we want to."

"It's a horrible name," said Daniel Briere, president of TeleChoice, a New Jersey telecommunications consulting firm. "The good news is that it doesn't sound like anything else; the bad news is there is a reason for it." While the name was somewhat controversial, the logo came under even greater attack. In designing the logo, Schacht had made a number of criteria clear. He wanted no "sunshine colors"; the logo should be in a primary color, but not AT&T blue. Landor scrapped the apricot-colored egg-shaped designs it was considering and began work on red logos. Soon the walls of a conference room were covered with potential designs for the executive team to view. The pictures included a render-ing of Alexander Graham Bell, arrows, circles, and versions of spirals. Some were hand drawn, a risky avenue the Landor people thought Lucent's management would reject.

At one viewing, Carly Fiorina looked at a cherry-red hand-painted circle that reminded her of her mother's paintings and expressed her strong preference for the unorthodox design. McGinn made it clear that he was not overjoyed with any of the choices but would go along with Fiorina. He felt that the design had a human quality to it that was both unfinished and unmachined, suggesting that the potential for human innovation is an unending process with an unlimited future. The final logo, a hand-drawn ring that resembles a lipstick kiss against a mirror, drew heavy criticism inside and outside the company. Dubbed the "innovation ring," some employees felt it looked like a red doughnut drawn by a small child, or worse, an advertisement for a paint company.

Marc Schweig worked for Network Systems Asia/Pacific and was in Japan when the new name was announced. He missed the employee broadcast, which for him happened in the middle of the night. "I remember running downstairs to the hotel lobby to get the fax. I wanted to know about the logo name and identity before I saw any customers first thing that morning," Schweig said. "Later, when I met with customers whose company was affiliated with a Japanese automobile company, I told them our new name and showed them a picture of the new logo. I tried to convey what the new company would be like and the important role Bell Labs research would play. But all the while I was speaking, they were snickering and I could hear the word 'doughnut.' I later found out that there is a tire company in Japan with a very similar logo and their tires are nicknamed 'doughnut.' "

Even Fitzgerald, who would be charged with putting on the company's public face, was less than thrilled: "It turns out you don't need to love the name or the logo to be able to turn it into one of the best known names in communications in less than two years. Because, trust me, I was at best lukewarm about the name—with its key virtue being that it wasn't a made-up name and was actually a word in the dictionary meaning 'marked by clarity and glowing with light.' And I hated the logo because it looks like an ink smudge and it's hard to duplicate." A number of journalists described it as a big, fat, red zero. Yet, as one Rutgers University professor who weighed in on the debate noted, if the company succeeded, the name would too.

On the face of it, Lucent's campaign to advertise its new name made no sense. Spending $100 million on television and print ads to raise the profile of a company with products that few people understood and even fewer were in a position to purchase did not seem like a promising investment. But customers were only a small part of the rationale for the hefty expenditure. In September, millions of investors would open their mail to find they possessed Lucent share certificates. If they had no knowledge of the company that stood behind that name, if they had to ask "Lucent who?," was there not a risk that investors would sell the stock in droves?

The advertising was also designed to create an image of the sassy, brash, fast-moving company that Lucent hoped to become. Changing the perception among employees and customers was central to the campaign. The ads had their intended effect. A survey of employees conducted while the initial campaign aired showed that the perception of their employer was altered with the ads. It speaks volumes about the power of advertising that those who worked at Lucent every day—perhaps had been employed there for most of their working lives—felt differently about their careers because of what they had seen on television at home the night before. One hundred million dollars was a steep price to pay, but Lucent's management believed that unless self-perception was altered, behavior would not change. This was the first step.

Behind the smiles and laughter at the February announcement, Schacht and McGinn's audience was nervous and skeptical. Theirs were the faces of Middle America. They had signed up to work for the telephone company, not some unnamed start-up whose management had just selected a name that means "glowing with light and marked by clarity" and then announced in the next sentence that the company "will be whatever we want it to be." The new leaders needed to win them over. McGinn tried reassurance; he was much closer in age and experience than Schacht to those he was trying to persuade. The company's president spoke without notes, letting his audience know that he understood their concerns and that he was one of them, mentioning his twenty-seven years with AT&T and referring to the spin-off as "leaving the family." He discussed his own misgivings about seeing his name on a business card without the AT&T logo and told his audience not to worry, that they should expect to have mixed emotions. "You don't get this feeling out of your system overnight," he told them. He was second-generation AT&T; there was nothing that they were feeling that he did not understand.

McGinn referred to Lucent as the "firm," as if it were some small family organization into which they had all been born. He announced, "We are no longer the people of AT&T. We are the people of Lucent," to roaring applause. He urged his audience to have fun. He wanted Lucent to be a place employees would be thrilled to be part of and entreated them to "blaze a new path that becomes known as the Lucent way of doing things." He did not want employees to think about where they had come from but rather where they were going. Schacht took a different approach. He was all business, with his buttoned-down manner and midwestern lexicon sprinkled with expressions like "wow, my goodness gracious." He was the father figure who explained the nuances of SEC regulation and the legal technicalities and time line they would all observe as AT&T and Lucent separated. Schacht reframed the issue, determined to show employees and investors how Lucent would be different. Lucent was not leaving AT&T; it was embarking on the "opportunity of a lifetime," a phrase that Schacht felt summed up the prospect that lay before them and one that he would use every chance he got.

The employee audience was focused on more concrete matters. The crowd donned white baseball caps, emblazoned with the Lucent logo, that clashed painfully with their conservative business attire (in one of the day's more touching moments, Schacht and McGinn later autographed every one of those caps while their employees waited in line for an hour). Then a runner dashed out of the auditorium to hoist the new red-and-white Lucent flag outside the Murray Hill headquarters, and a cascade of red and white balloons rained down. Management asked the audience for their thoughts. As the meeting progressed, Kathy Fitzgerald read aloud the questions that were pouring into her fax machine from locales all over the globe: What will be the policy on pensions? How will salary grades change? How should we answer the telephone?

During the long presentation, McGinn shared the credit for the creation of this new company with his "world class" management team. He lauded their efforts as nothing short of superhuman and asked each member to rise and take a bow, without mentioning that many had strenuously objected to joining Lucent and that only his personal pleas had convinced many of the team to sign on in the first place. When AT&T had spun off Lucent, it bundled together all of the businesses that roughly fell under the rubric of manufacturing and sent along a few other divisions it no longer wanted. None of these businesses, with the exception of Network Systems, wanted to be part of Lucent. Network Systems had been a world unto itself, a distinct subculture within the AT&T universe that not one of the other departments wanted to be subsumed by. Many viewed Network Systems, dressed up with its 1990s moniker, as nothing but the old Western Electric manufacturing company. Underneath they saw it for what it was, an almost solidly white male organization where the factory manager was king. Fiorina remembers that most people from the more elite Long Lines business were underwhelmed by the equipment makers: "The rap on Network Systems was that it was all guys with 20-inch necks and pea-sized brains. You know, heavy metal bending. I went because it was a huge challenge, completely male dominated and outside everything I'd experienced."

Lucent, liberated from AT&T, was a portion of Bell Laboratories and a conglomeration of several ill-fitting businesses barely held together by the common theme of manufacturing. Part of the rationale for melding the disparate enterprises was that Lucent's structure and product mix would mirror that of many of its competitors. The more salient argument was that it suited AT&T's purposes to dispose of extraneous and less profitable departments. Allen wanted to take AT&T entirely out of the manufacturing business, and so, in spinning off Lucent, he sent with it all of the businesses that fell into this category. The primary businesses set free included switching, microelectronics, business communications systems, and the consumer phone businesses. In essence, AT&T bundled up those businesses that fit with Network Systems, along with a few others for which it had no use, and foisted them upon the new management. Lucent thought of its core businesses as systems, silicon, software, and services. Yet thrown into the mix was one company that assembled modems, another that manufactured printed circuit boards, and a third that sold telephone handsets. Schacht and McGinn made it clear that they did not want these businesses, but neither did AT&T, and it called the shots.

Ideally, Lucent would have immediately rationalized this new structure by selling those businesses that were either unrelated or lossmaking, but it was constrained by the terms of the spin-off. For three years Lucent could not dispose of any assets worth in excess of 10 percent of the company without causing AT&T to lose the tax benefits of the spin-off, which in turn it would have simply billed to Lucent. As one senior executive remembers, "We were stuck with anything that was major regardless of the strategic or financial value." Lucent had to try to make the structure work, knowing that as soon as it was given a chance, it too would be liberating some divisions.

In the early days there was much public discussion about the synergies that would result from throwing together these previously unrelated businesses, but Lucent's own management was skeptical and the arrangement ultimately proved unsustainable. Curtis Crawford, the head of Microelectronics, and Pat Russo, head of Business Communications Systems (BCS), did not want to be part of Lucent. Both executives went to Schacht and McGinn and argued that their businesses should be spun off from AT&T as independent companies. They would be two of the most senior executives at Lucent, and before they even got in they were trying to get out. McGinn and Schacht had no leeway; the shape of Lucent had been determined, and now each executive had a choice to make. McGinn sat down individually with his nine most senior managers and gave them an option. He told them, "If you come to Lucent you need to personally commit to this as a positive change. You need to come because you want to, not because you are just following a job." This was a new company, not just a new organization chart, and he wanted a personal commitment from each member of his senior team. One by one, over the next few weeks each of his most senior managers came to McGinn with a pledge, although a number of those in second-level management positions found their way back to AT&T.

Network Systems was and would remain Lucent's largest and most important division, with 55 percent of the company's revenues and 52,700 employees. This business entails designing, building, installing, and servicing all aspects of the communications systems (voice, video, and data) for large network operators, including local, long-distance, and wireless telephone carriers. One of Lucent's most critical competitive advantages was that, in 1995, almost 60 percent of all phone lines in the United States were already hooked up to one of its switches. Lucent's equipment dominated every aspect of the U.S. telephone networks, and its relationship with the incumbent carriers dated back generations. Most of Lucent's top management came from this division, and the focus of the entire corporation, at least initially, would be on Network Systems' most important customers, the Baby Bells and AT&T. They were the customers for which Bell Labs focused its research and product development and with which the sales force, as former colleagues, felt the most comfortable. At the time of the spin-off, Network Systems had a dominant market position in almost every business in which it competed in the United States, and where it was not number one, it was a close number two. With more than half of the new company's revenues, Network Systems would remain Lucent's primary business after a long series of sales and spin-offs eventually excised other businesses.

The Business Communications Systems division provided private branch exchange telephone systems (PBXs) to small, medium, and large businesses (known in the industry as "enterprise customers") that were capable of switching calls within a company and externally. Here, too, Lucent had the largest installed base. With 26,000 employees, BCS was the second-largest division within Lucent, contributing more than 20 percent of the company's revenues. BCS was run by Pat Russo, a tall, slender, athletic woman who gives the appearance of great self-discipline, although she readily admits to a chocolate addiction. Russo was by all accounts a star at AT&T, and the spin-off offered her very little. Although publicly supportive of Lucent in every way, privately Russo had hoped for a reprieve. Her ambition had nothing to do with Lucent, and she hoped that Allen would make her CEO of a new company that would be spun off separately and sell data (merged perhaps with competitor Bay Networks) and voice (BCS) systems directly to enterprise customers. But a deal with Bay Networks never materialized, and BCS was thrown into the Lucent mix. If Russo was unexcited about joining Network Systems, Network Systems was less than enthusiastic about joining forces with BCS, as few anticipated much growth from this mature business that throughout Lucent's history was always in the process of being fixed.

Publicly, Lucent's management sang the praises of its Consumer Products group, highlighting the fact that Lucent's products still hung on the kitchen walls in the majority of American households. In truth, it was a business beset by obsolete product lines and revenues in terminal decline. Selling telephones and answering machines at small outlets all over the country did not mesh with management's plans for a high-tech research and manufacturing company. With 6,700 employees, Consumer Products was the smallest business to depart with Lucent, and while Lucent continued to make public statements about the ambitious plans it had to return this division to profitability, revenues declined by 25 percent in Lucent's first year. The business did not make fundamental sense for Lucent. The outlay that would be needed to keep a consumer brand before the eyes of the public would run into hundreds of millions a year; Lucent had no experience running a retail chain; and the company would have a difficult time retaining any advantage over low-cost competitors.

Schacht gave Fiorina the job of president of the Consumer Products division. Fiorina's mandate was to overhaul the product line and build on the strengths of the retail franchise. Instead, in 1997 she announced a more decisive action and merged the business with the Dutch conglomerate Philips into a new endeavor called Philips Consumer Communications, in which Lucent would hold a 40 percent interest. The jointly owned venture was a $2.5 billion company that sold telephones and answering machines and would have no more success than Lucent had had on its own. It began life as a money-losing entity, with slow product development, and nothing changed.

Schacht, however, would hail her action as revolutionary. "Remember, he said, "this is the company that invented the telephone, so the idea of giving up that business wasn't too obvious to any of us at the time. Carly made an absolutely correct decision. And she did it without knowing what her next job would be." But the press was more critical of her decision: "As president of Lucent's consumer-products division two years ago, she decided to form a joint venture with Dutch electronics giant Philips Electronics NV to try to resuscitate Lucent's consumer wireless and phone business. The venture failed so badly it was discontinued last year, and Lucent is selling the businesses. [Fiorina] has called the Philips venture the biggest mistake of her career," *The Wall Street Journal* observed.

The microelectronics business, while technically a manufacturing business and therefore aligned with Network Systems, again operated separately from the Network Systems business and sometimes at odds with it. Although initially unpromising, the business of developing and manufacturing integrated circuits and optoelectronic components for use in telecommunications was about to experience tremendous growth. However, in a short time it would be apparent that microelectronics would have the same conflicts operating inside Lucent that Lucent had as a division of AT&T. Nortel, Siemens, Ericsson, and other phone suppliers were reluctantly customers of Lucent's microelectronics business only when they could not get the supply someplace else. It was another arrangement that could not last.

Despite the auspicious market conditions Lucent sailed into from its earliest days, the litany of strikes against the new company was daunting. Lucent began life with a president who thought he should be CEO, senior managers who had not wanted to join the business, customers that still seethed with long-held resentments, a culture that would need to be reinvented, an unsupportable cost structure, and a workforce that ranged from highly enthusiastic to angry and resentful at being abandoned to their fate.

CUTTING THE CORD

"Initial public offering," or IPO, a once-obscure financial term, slipped easily into the lexicon the summer of 1995 when Netscape, which went public at $28 and on the first afternoon traded at $75, showed the world what could happen to a newborn company if it used the word "Internet" or "network" in its offering document. IPOs had once held an element of risk: the company was unknown, its business might flourish or fail, and the stock price could trade in either direction. For years, investment banks refused to take companies to market that had not shown sustained profitability. When Netscape went public, few outside of the investment or high-tech worlds could have described its product or how to use an Internet browser. But in the months and years following the offering, something fundamental changed. It was no longer necessary to know much about a company in order to make a quick killing on its stock. If one took the time to learn about a company, study its product, its market share, its balance sheet, and its prospects for the given industry (which were almost certainly unknowable), the opportunity would have passed. IPOs became get-rich-quick schemes, doled out to those who had close ties to the underwriters, and the question that hung over each issue was not whether it would make money, but simply how much.

In February 1996, Carly Fiorina made her public debut as the coordinator of Lucent's IPO, a position that immediately brought her to the attention of the financial community. She had risen steadily at AT&T and was the first female executive vice president in Network Systems. The success of the IPO resulted in further promotions, and she was later named group president of Lucent's global services provider business, accountable for more than 60 percent of Lucent's revenues. Fiorina did not have a typical AT&T background. A graduate of Stanford University in medieval history and philosophy, she had dropped out of law school, a difficult decision for a lawyer's daughter. She says, "I wanted desperately to make my father proud that his daughter would follow in his footsteps. Quitting was the ultimate personal failure in my mind and his. Yet, in the end, loving what I did was more important. But life went on." Fiorina then embarked on a series of jobs, including receptionist and teacher, before she ended up at AT&T as a sales representative in 1980 selling phone systems to government agencies—not an obvious fast-track position. Initially, she remained so unconvinced about her tenure at AT&T that she refused to join the savings plan.

Yet it was at AT&T that Fiorina came to truly love a challenge. After divestiture in 1984, she gravitated toward one of the company's most troubled and perhaps boring divisions. Access services was responsible for connecting long-distance calls to the newly independent local phone companies. Putting systems in place was a morass into which Fiorina flung herself. "This is not something that most people would think of as fun," she notes. "People thought I was nuts. Access Management was an area about which I knew absolutely nothing. Nevertheless, our goal became to verify every bill and prove every overcharge. I've saved a picture from those days where I'm standing in a room covered floor to ceiling with boxes filled with bills. A team of us looked over every one of those bills manually for three or four months and found significant overcharges."

Fifteen years after walking through the doors of AT&T, Fiorina was asked by McGinn to coordinate the largest IPO in history. Her vision of Lucent mirrored his own. "Lucent is about what happens when you see the possibilities, not the limitations," Fiorina said. She compared the Lucent spin-off to the splitting of the atom, an event that released a mountain of energy. When he asked her to orchestrate the offering, McGinn not only expressed his confidence, he raised her from the pack of executives. "Carly is wickedly smart," McGinn says. "I told her that given her knowledge of the outside world and her ability to synthesize disparate flows of information, no one could do the job better." McGinn's only reservation was Fiorina's lack of experience with underwriting.

It was an enormous responsibility. AT&T had a long history of successful offerings and was determined that this one would not fail. But Jeffrey Williams, a Morgan Stanley investment banker for the deal, recalls, "She had little experience with finance. She was learning while doing. But Carly is so bright. She was always urging us to think in new ways about positioning the company to investors." Fiorina stood out a mile from the gray suits of AT&T, and soon her demeanor came to personify Lucent. Fitzgerald remembers Fiorina clearly: "Her clothes were at the high end of style: a very form-fitting black top, a very stylish skirt. She wasn't flirtatious, but she had an enticing personality. It was just incredible to have a female executive who was that attractive, that young, [and] that effective." But Fiorina brought far more than a great wardrobe and practiced presentations. She brought fun.

Initially, AT&T's bankers tried to temper expectations for the new stock. Equiptco, as Williams and his Morgan Stanley team liked to call their client, would start life with a huge installed base, which would be an enormous advantage for future sales. But the new company's growth prospects were considered dim at best. Its recent growth had been in the single digits, it lacked well-known management, and over the previous five years it had accumulated losses in excess of $5 billion. Morgan Stanley warned Lucent's management that the price would almost certainly decline after the IPO because of the supply of stock that overhung the market with the upcoming spin-off. Originally it told the AT&T board that the offering price would be between $17 and $19 and that the board should understand that during a digestion period many people would dump the stock. The board could expect the price to languish in the $14 range for most of the summer and perhaps rise back to the offering price by the fall. After a few months of foundering at perhaps 25 percent below the offering price, Lucent could reasonably expect the shares to return to the offering level, but not much higher. It confidently informed the AT&T board that what they were selling was hardly a hot property. "People thought we would sink like a stone," Fiorina remembers.

For those who studied Lucent's finances closely, there was nothing about which to get excited. In 1995, the telecommunications sector was still viewed as a moderate growth business, with some healthy prospects overseas, where Lucent's competitors were better placed. The industry traded at eighteen times earnings, and it was expected that Lucent would lag this multiple by a couple of points. Wall Street analysts remained underwhelmed. Most had not yet grasped the magnitude of changes in the telecommunications industry that would sweep Lucent up and support its astounding growth. In the first reports after the twenty-five-day quiet period immediately after the IPO (in which analysts were enjoined from making public remarks about the company), analysts gave Lucent weak "buy" recommendations, vastly underestimating the growth in the networking market. Their concerns were manifold, starting with the management. Top talent, it was agreed, was still at AT&T. Network Systems' managers were considered by the company to be second-stringers. But despite a loss in 1995, a boom in new business from the Baby Bells once the separation was announced improved Lucent's financial results for the fourth quarter. And while it was widely believed that Lucent was a captive supplier to AT&T, when prospective investors learned how little business the two companies actually did together, Lucent's outlook appeared brighter.

Then there was the culture. All of the familiar traits were thought to apply: slow, bureaucratic, risk averse, and stymied by a long history of cost-based-rate-of-return thinking. Most believed that Lucent would struggle when taking on the fleeter-footed competition. Sure, Lucent had a number of undeniable strengths. It was the industry leader, but the competition was gaining fast, as small cutting-edge companies like Ciena entered its markets. Ciena had shocked Lucent in 1996 with sales of a dense wavelength division multiplexing (DWDM) device that vastly increased the amount of traffic that could be sent through a strand of fiber-optic cable over any product Lucent had commercially available. Lucent had led the fiber-optic business; Bell Labs had invented the device that converted electrons to photons and made it possible to use lasers to send information over strands of glass. Lucent rushed to get its competing product out. Would a company carved from AT&T be able to meet this challenge?

In March, only a month after the new company announced its name, Lucent's road show began in London. On a whirlwind tour when Lucent executives would make their case to investors all over the globe, Schacht, McGinn, and Fiorina faced 175 curious investors over breakfast at the Savoy Hotel. Fiorina had tried to package Lucent as a technology-rich industrial company that could both invent and develop cutting-edge products. While investors accepted this vision, they flooded the new management with questions. With no technology background, how would Schacht run the company? Was the spin-off irretrievable, or would investors find themselves minority shareholders in an AT&T venture? Would the compensation structure for executives be appropriate for a high-tech company? And ironically, given the furor that would surround the subject of executive compensation only a few years later, was management being given enough stock options to create proper incentives?

For the IPO road show, Lucent's top nine executives split into three teams to cover the world. Schacht viewed the road show as one more opportunity to cement the notion of a united company. Each presenter now had to be conversant with the whole company, and in the course of their travels the group began to coalesce. In front of a group of initially skeptical investors, Lucent's executives could hardly refer to other parts of the company as "them" and thus melded into Schacht's notion of "we." In a trip like none of them had ever taken—thirty-four cities in fifteen days—they pitched their message to an increasingly enthusiastic audience. Later they compared stories of lost luggage, missed flights, and nasty meals, and the experience further bound the group together. At one stop late in the afternoon, Schacht asked Fiorina if she had even had time for lunch. She told him the packed schedule had not given her a free moment. Then the CEO went out, crossed the street, and found her a sandwich. As the bankers watched out the window, they noticed that Schacht had stopped. He had noticed a truck with a new diesel engine, and as Cummins business would always be part of him, he stopped the driver and asked what he thought of the truck.

At one midwestern stop, controller Jim Lusk and Fiorina discovered when they landed that they had no car waiting. Fiorina spoke to the pilot, who offered to lend them a friend's car. When they located the vehicle, they found it a wreck, an old car with torn-up upholstery. Fiorina only laughed, quickly writing out a sign to place in the window, "This is what cost cutting is all about." Lusk and Fiorina photographed the car with its message and brought the picture back to their colleagues in New Jersey to reassure anyone who had doubts about the pair's ability to economize.

The traveling squads were highly competitive as each tried to score "sales credits" (not actual money but acknowledgment) for their presentations. After each investor meeting, they would hound their bankers for a "score" of how many orders for the stock their meeting had yielded. When a few of the larger investors hosted multiple presentations from the various teams, there was a good deal of friendly bickering about where the credit for these new sales should be tallied.

Although it was a different city and a different audience every time the Lucent teams looked up, the message was always the same, and soon it took on a stale, repetitive feel. Scripts were followed too closely and pat phrases appeared and reappeared, boring the team members who were forced to listen in. To combat battle fatigue, Fiorina and banker Williams designed a game that would sharpen the speeches and stoke the team's interest.

Each day one of the well-worn turns of phrases would be banned for the day. If the speaker slipped into habit and used the offending term, he or she would make a nominal donation to a kitty, something on the order of five dollars, and at the end of the day the speaker who had made the fewest contributions would win the kitty. Although highly amusing, the game did not last for long. Schacht thought that his team was too distracted by carefully crafting its words and called it off.

Schacht soon got used to Fiorina's style. On an AT&T corporate jet flying back from Europe overnight, Lusk and Fiorina played Simon and

Garfunkel tapes and sang along, despite the fact that Schacht was trying to sleep in a chair nearby. First they played "Bridge over Troubled Water," but no, they decided, too somber. Then they blasted "The 59th Street Bridge Song." And as they belted out choruses of "Feeling groovy/Life I love you/All is groovy," Schacht squinted over at them, gave them a reluctant smile, and said, "Come on, give me a break."

Lusk believes that Fiorina's enthusiasm was one of the factors that convinced investors that Lucent was not an ossified group of phone company executives. "She [got] so psyched in these meetings. It's like the Super Bowl or the Emmys or something," Lusk recalled. "She's not cranked up on coffee or chocolate or anything but she's cranked up on energy. No one does [presentations] like Carly." Lusk knew that Fiorina's commitment was far more than showmanship. The night Lusk spent at the printers in downtown Manhattan trying to complete the final offering prospectus became a fiasco. In pitched battles with a battalion of AT&T lawyers, Lusk was losing ground. They were fighting over the fine print, but Lusk knew that there were details Lucent would later have to live with. At 3 A.M. he called to wake Fiorina, telling her that he could not get the lawyers to budge and needed her help. She drove from deep in New Jersey to be with Lusk by six. "That's how much she cared about this thing," said Lusk. "She would never ask anybody to do anything that she wouldn't do."

At home, the bankers pitched the stock to their own sales force, and each morning underwriter Morgan Stanley would have a "Lucent Moment" when a new Lucent factoid would be announced over the public address system. Morgan Stanley's sales force was regaled with Lucent minutiae: how many patents per day came out of Bell Labs, how many Nobel laureates were employed by the new company, the average number of years of education each member of the research staff had under his belt. With or without the factoids, selling Lucent was not proving to be difficult. To investors, Lucent looked like the holy grail of investments: a low-risk, high-tech company. It was an anomaly among other high-tech offerings: a company that had real revenues and customers. It had the distinct appeal of a proven product that investors understood. Stockbrokers hawked the shares with the slogan Lucent had spent $100 million to publicize: "We make things that make communications work."

The New York investor luncheon held at the Plaza Hotel was scheduled for halfway through the trip. At each stop the crowds had grown, like a rock concert tour gathering groupies. Those who watched Schacht, McGinn, and Fiorina speak, who had some knowledge of the arcane dance played out between the vendors and buyers of public companies, saw the performance of Lucent's top managers improve with each stop as they refined their message based on Fiorina's acute instinct for selling. Five hundred investors attended the New York show as the team was at its peak performance. Schacht conveyed sincerity; he called Bell Labs a $2.1 billion idea machine, "the competitive advantage we take into the marketplace." And as the china clinked and the glasses were raised, Schacht told his audience, "Given a chance to form this partnership with Rich and his merry band of achievers is something I could not pass up. They are young, they are aggressive, and simply something I could not be without."

Fiorina had given the entire show energy and dash, and as investors ate their lunch she further whetted their appetites by telling them Lucent was no pie-in-the-sky opportunity but a company staring at $100 billion worth of real-growth opportunity. "Forget sexy services," she told them. "Forget interactive TV. Forget the Internet and all that. We, fundamentally in this country, have an access network that needs to be replaced." She closed the presentation with "If you like our track record, you need to watch us now." Even some of the jaded New York investment types, people who watch these presentations for a living, rose to their feet to applaud her performance. That night the bankers canceled the remainder of the road trip; the demand for the stock was so great that they priced the issue early the following day at $27 a share without bothering to finish the rest of the tour. Although the offering was for only 112 million shares, or 17.6 percent of the company, the underwriting syndicate already held requests for 735 million shares. The offering netted AT&T $3.3 billion. It scheduled the spin-off of the 523 million remaining shares to its shareholders for September 30. Banker Jeffrey Williams, a road show veteran, remembers Lucent's as the most successful road show ever.

Schacht and McGinn viewed the IPO as an opportunity to bequeath every employee a block of shares. They believed this action would not only help foster an owner-employee outlook but also crucially further the "we" mentality they were working so hard to create.

The bankers had indicated early on that they could hold back a block of stock for the new company's employees, as they did not expect a stampede. Suddenly, just days before the listing, AT&T backpedaled. The notion of handing over shares to what amounted to departing employees became a less interesting notion to the seller, AT&T, in the face of burgeoning investor demand. Schacht was livid. Years later McGinn was still furious that they had not been able to make this grant to their new employees, but their hands were tied. This was not their IPO but AT&T's.

In the end, the gift was replaced with stock options for one hundred shares for each of the new company's 131,000 employees. (At their peak, each employee's options would be worth roughly $20,000. Up and down the New Jersey shore in the summer, boats could be seen with names like "Lu at 72" and "Lu at 85" emblazoned on their hulls.) But options shifted the burden of the cost of this gift from AT&T to Lucent. The day before the offering, the investment bankers returned to Schacht and said that AT&T had set aside enough shares for grants to the top fifty managers. Schacht presented this largesse to his team and was thrilled with the response. Every single one of the senior managers turned down the offer. They felt that it would be unfair to take something AT&T would not offer to employees. Schacht felt that this was a very early indication that he and McGinn had created something quite different. Lucent's executives received no shares at the IPO price, even though their AT&T counterparts did. For both symbolism and substance, this gesture made an enormous impression on Lucent's employees, who now saw that the "we" thing that Schacht had cited again and again was more than just talk.

On April 4, 1996, after six months of intensive preparation, Lucent was free. By tradition, the CEO of the new company should have been the first to purchase shares in the newly floated company, but Fiorina's planning extended to both the substantive and the symbolic. Nellie Perkins, a curly-haired production employee from an Oklahoma City manufacturing plant, had her fifteen minutes of fame when she rang the opening bell and bought the first share of Lucent on the New York Stock

Exchange floor. Later, as she was jostled around in the melee on the exchange floor, she gleefully told the news cameras, "It's not like Oklahoma." Lucent employees from all over the world were flown in to New York to be at the stock exchange on the first morning of trading, and the rest could watch the show on an internal employee broadcast. The day ended with Fiorina and Lusk ringing the closing bell from the podium of the stock exchange and Fiorina telling reporters, "I fell like I've just given birth here, literally."

Schacht and McGinn were worried that the uncertainty of the IPO might cause customers to seek additional, or even alternative, suppliers for their products, so immediately after the investor road show they embarked on a customer road show that covered the globe. They tried to drive home the message that service would improve and product supplies would be uninterrupted. The local phone companies were stunned by this very un-AT&T approach. "We can't believe the attention we're getting now from the new Lucent," said Raymond Smith, Bell Atlantic chairman, who had a reputation as an AT&T basher. "The [Lucent] CEO called and asked to meet me—the first time an AT&T CEO had done such a thing."

Lucent investors, the legions of former AT&T shareholders, displayed far more confidence in the new company than the Wall Street experts. This vast investing public, who otherwise might never have participated in the tech stock run-up, were suddenly holding, and profiting from, one of the new bellwethers. These same investors might never touch Cisco, Amazon, or JDS Uniphase, deeming them to be highly risky investment vehicles, but they gladly held on to their Lucent. After all, it was Ma Bell. Part of the credit has to lie with the highly successful and engaging media campaign Fiorina ran in the months before the new share certificates arrived in the mail. But perhaps the reason shareholders held Lucent is far simpler. By the time the AT&T board distributed the remaining shares to grateful shareholders six months later, the price had risen by 70 percent. For Lucent's management the IPO was a triumph, the culmination of six months of intensive effort that had begun the moment Allen had announced his decision to let them go.

THE BATTLE WITHIN: LUCENT'S CULTURE

The biggest challenge Lucent faced was not Nortel, Cisco, or even satiating Wall Street's appetite for growth. Lucent's biggest test was its own culture. Top Lucent management acknowledged that even eleven years after divestiture, AT&T had not moved far enough from its monopoly legacy, from the days when competitive challenges were met in the courtroom rather than the marketplace. Almost all of Lucent's management had begun their careers with AT&T and thrived in the halcyon predivestiture days. They were all too aware of AT&T's less-than-flattering reputation and could not help but feel slightly defensive. McGinn once remarked, "We're not a calcified group of people with some secret handshake."

Lucent would urgently need to reinvent its culture, shedding much of what it had lived with under AT&T. And it would need to make this transformation at a time when its industry was experiencing its greatest turmoil and growth. While there is no doubt that in separating from AT&T, Lucent successfully managed to create a new image (by no means an inconsequential task), the essential question was whether it would be able to create a new culture as well. With the help of the advertising agency, Fiorina had created a series of ads that were revolutionary in their simplicity. A disembodied male voice, speaking over the tapping sound of a computer keyboard, reads the messages he types, which the viewer can read on the screen:

> Let's see . . . About six billion people on Earth.
> Less than half with phones.
> Most without cellular.
> Even fewer on Internet.
> Just discovered life on Mars.
> Got work to do!
> Lucent Technologies. We make the things that make communications work.

On Lucent's first day as a new company, Fiorina dressed in a Lucent-red suit and told all the company's employees that the campaign had already been an overwhelming success. "We set out to make our advertising unique and memorable, and I think we've accomplished that," she announced. "Our personality, as revealed in our advertising, is engaging and endearing," then, nodding at her bosses on the stage, "like Henry and Rich, like all of us. He's funny. He's informal, entrepreneurial, confident, but not cocky, with a little bit of an attitude, you may remember. And in answer to the most asked question, 'No, you will never see his face.' " She told them with pride about meeting Senator Bob Kerrey of Nebraska only weeks earlier, whose first comment to her was, "Love your advertising." The new image certainly helped Lucent attract top talent, garner customers, and reenergize its own staff, but the advantage would be sustained only if the transformation were more than skin deep.

Neither McGinn nor Schacht underestimated the test Lucent faced over the coming months and years. The distance Lucent would need to travel from its slow-moving heritage to the technologically advanced world was greater than most companies travel in decades. Schacht presented a worldview that was nothing less than a revolution from the days at AT&T. First, he told Lucent employees, "Technology is going to create and obliterate markets and products overnight. Developments in faraway places will have immediate impact. Markets will expand and contract violently. Functional skills will become obsolete without warning." For employees of a company with lifetime employment and products that lasted a generation, his words were a wake-up call.

AT&T was once a world of its own, with a million employees and a century of history, characterized by a conservative outlook, reliability, and a deep respect for the power of technology. AT&T stood for durability; if it made a product, it would continue to support it for the life of the product no matter how obsolete and cost-ineffective such backup might become. AT&T would not junk the product, make a 90-degree turn, and ask customers to start again with something new. At AT&T there were few surprises. It provided employees with stability and prestige, and they returned the favor with long-term loyalty, sometimes stretching across the generations.

Those who spent many years with AT&T describe its inward focus, the obsessive concern with internal positions and the life of the company, rather than any focus on the outside world. "I probably worked just as hard at AT&T as I do now. It was just on the wrong things: the power struggles, the budget debates, the political issues," said Alex Mandl, erstwhile president and COO of AT&T and former CEO of the CLEC Teligent. The company had an inflexible hierarchy, and one former president of an AT&T unit remembers that employees used to introduce themselves to each other using both their name and management level. Confidence and self-regard verged on insularity and arrogance. There was a common belief that only those who had earned their stripes inside AT&T would ever understand the complex world of telephony. AT&T drove the communications market, not the other way around. Innovation ran on its schedule, and product cycles could last as long as twenty-five years. Market share had always been AT&T's to lose, so it never had to pick up the pace. Nothing could have been worse preparation for the competitive world in which AT&T—and Lucent—would find themselves by the 1990s.

Although loath to publicly disparage their mutual past, as AT&T remained its largest customer and an institution to which its own employees remained steadfastly loyal, Lucent's management had to point out the shortcomings of AT&T and the ways in which Lucent would change. Otherwise it could expect only more of the same. In the first all-employee broadcast announcing Lucent's name in February 1996, Schacht walked this difficult line. AT&T, he told his audience, had many strengths. But Lucent employees needed to ask themselves what about the 127-year-old traditions they held dear and true. The list of qualities that Schacht felt were worth clinging to included respect, integrity, staying power, reliability, and stature. And carefully, without pointing any fingers, he suggested that Lucent was an empty vessel that needed to be filled with new qualities such as speed, flexibility, competitiveness, and a customer focus, which, by implication, had been missing.

Schacht and McGinn changed the tone of the company dramatically through some simple but powerful gestures. They ate in the company cafeteria, never bothering to build an executive dining room. The standard corporate offices, with their light-colored wood-and-glass fronts, are all of a similar size. Schacht's and McGinn's offices could be reached by simply pushing "3" on the central elevator bank and walking down the hall. At AT&T, the top executive offices were behind locked glass doors and accessed by an internal spiral staircase that only the select were allowed to ascend. Both men wandered the company in shirtsleeves, and one longtime AT&T/Lucent product manager remembers being struck by the fact that "Henry Schacht would talk to anyone." One sales manager remembers arriving early for a meeting at Lucent's Berkeley Heights office and wandering into the cafeteria only to find Schacht standing in the cashier's line holding two briefcases, a bran muffin, and a cup of coffee. For employees, most of whom had never even caught sight of their top bosses at AT&T, this was earth-shattering.

To many of those who spent their entire careers with AT&T, the transformation was profound and entirely welcome—the signs of a company keeping up with the times. Lucent made decisions more quickly, integrated engineers and marketers on development projects, and distributed stock options. For the first time, the company focused squarely on the financial markets and marched to the beat of the "shareholder value" drum.

Yet there were intense frustrations, and Lucent bravely probed these further, looking for real information rather than simply anecdotes. In 1997, Lucent sent questionnaires to its 122,000 employees soliciting their views of the company. Many of the questions had been used in an AT&T survey, but Lucent pointedly dropped the question that focused on employment security, as it was "no longer unconditionally expected or given." In a gutsy move, management compared Lucent's results to both the industry's and the high-performance companies' it was explicitly trying to emulate, and then published its findings. The responses were telling: the company had moved a great distance, but for many it was not nearly far enough. Lucent employees loved working for the new company and saw job satisfaction, learning and development, and the company's competitive position as strong points. But, crucially, innovation and risk taking were regarded as the weakest areas. More than half felt that decisions were still made too slowly and that the company was inefficient and inadequately supported the risk takers in their midst.

When Lucent compared its culture to those of a group of high-performance companies with ambitious financial goals and strong reputations (a group that included American Express, Hallmark,

Hewlett-Packard, Nestlé, and RJR Nabisco), the weakness showed up in innovation, speed, and customer focus. Lucent trailed the high-performance cohort. Tellingly, less than half of the respondents believed that Lucent was "quick to take advantage of new ideas or opportunities." Pat Russo acknowledged the company's weak spots and suggested that the employee input would allow management to refine the steps that would move the company farther along. Management acknowledged that the questionnaire had increased the expectations for change and did not proffer excuses for the findings. Comments ranged from the enthusiastic "Go Lucent! I want to help my company grow" to the more critical "It is difficult to prove the value of a new project that requires innovation and risk-taking for success."

Those who came to Lucent from other organizations were stunned by what they saw. Fiorina had transformed its image from that of a telephone company castoff to a Cisco competitor. Lucent's financial results seemed to bear out this transition. But once inside, newcomers found a different organization. They said that to understand Lucent it is essential to understand AT&T, that Lucent was the apple that fell from the tree nearby. Many perceived an organization mired in bureaucracy with little ability to move at pace, in part because of its love of complexity. A top executive who had worked for a competitor described coming to Lucent as changing time zones. Life was now run on Parsippany Standard Time, a leisurely pace at which no one from, say, California could function. New Jersey was now the center of the universe, and internal politics often overwhelmed action.

The question of Lucent's leadership still loomed over the company. That McGinn would be given the reins of power was an unquestioned outcome, but the timing of the handoff still hung in the balance. On September 20, 1996, just ten days before the final spin-off of shares to AT&T stockholders, McGinn approached Schacht. He apologized for being early, as it was still a few days short of the one-year anniversary of their first leadership conversation. Schacht and McGinn sat down at the negotiating table. McGinn's public profile had risen with the IPO, and his haste to discuss the transition had been triggered by enticing job offers that were coming his way. A small, cash-rich company had recently made him an attractive offer that he was seriously considering, and he laid this option before Schacht to discuss. He viewed Schacht as a mentor and told him he wanted to talk through with him the pros and cons of leading a much smaller institution. McGinn was still in his forties and, without explicitly threatening Schacht or the board with his departure, simply said he needed to make a career decision. He told Schacht that he was looking for a sense of his timetable and the thinking of the board on this matter. Schacht told McGinn to give him a few weeks and he would have an answer.

The Rich McGinn who sat before Schacht in the fall of 1996 was an entirely different, and, in the world of headhunters, far more valuable commodity than the man who had occupied the same seat a year earlier. This fact had not escaped the board's notice. McGinn would have been a moderately interesting hire as the head of a division of AT&T, but as the CEO-elect of an already successful high-tech company, he was now a hot property. As promised, Schacht raised the issue with the board and in October returned to McGinn to tell him a timetable was in place. Schacht would leave no later than March 1998 and perhaps as early as October 1997. Without cutting off any of his options, McGinn thanked Schacht for the vote of confidence and told him that was the information he needed.

BELL LABS INNOVATION

Waxing lyrical about Bell Labs' vaunted past is a favorite activity among AT&T/Lucent watchers. These nostalgic fans never tire of reminiscing about the fact that "the Labs" were the birthplace of most of the major technologies that transformed twentieth-century life. Bell Labs, these enthusiasts will say, invented the cutting edge. Much of this nostalgia is grounded in fact. It is hard not to be amazed that in one small corner of New Jersey so many true innovations sprung to life, among them the Telstar satellite, the UNIX operating system, cellular telephony, and high-definition television, as well as radar, the transistor, and the laser. It is impossible not to believe that there is something profound about an institution that fosters eleven Nobel Prize winners for far-reaching discoveries from methods for cooling and trapping atoms with laser light to detecting the background radiation left over from the "Big

Bang." Few conversations about AT&T's history are devoid of descriptions of Bell Labs as a national treasure.

Bell Systems' first attempt to establish a research laboratory in the 1880s was aborted when its director, Hammond Hayes, ominously said, "I have determined for the future to abandon theoretical work for this department, devoting all of our attention to practical development of instruments and apparatus. . . . I think the theoretical work can be accomplished quite as well and more economically by collaboration with the students of the Massachusetts Institute of Technology and . . . Harvard College." The nascent organization had already stumbled upon the thorniest issue that would plague research at Bell Labs throughout its history. Striking the uneasy balance between commerce and science would be no simpler over the intervening 120 years.

While Hayes might have thought that he had solved this dilemma for his generation, a passion for exploration and invention became deeply embedded in the psyche of what would become AT&T and Lucent. Whatever commercial success AT&T/Lucent found, there is no greater source of pride for those who have worked for these two companies than the intellectual achievements of Bell Labs. Lucent's cavernous eight-story lobby, with its brown brick floors and vaulted concrete ceilings, is a small museum dedicated to memorializing these paramount accomplishments. Inside the pyramid-shaped atrium is a bust of Alexander Graham Bell that sits seven feet high surrounded by museum-style glass cases full of displays and models highlighting Bell Labs' achievements decade by decade. The Nobel Prizes are each commemorated with a plaque, and a model of the Telstar satellite swings from the ceiling. In front of the interactive displays is a large LED counter constantly tabulating the number of patents—in excess of 27,000—Bell Labs has garnered over the past three-quarters of a century. But Nobel Prizes did not pay the bills, and patents did not necessarily turn into products.

When Bell Labs first opened its doors on January 1, 1925, the telephone was fifty years old and there were 17 million users, 12 million of them in the Bell System. From then until divestiture, first in Manhattan and later in Murray Hill, New Jersey, Bell Labs thrived under the flood of funding that the telephone monopoly produced through an R&D tax levied on every phone call and telephone rental. As one former researcher described it, Bell Labs was "the higher brain function for the Bell System." The research unit derived its enormous power from the fact that it could veto any technological change in the country's phone system. Researchers in Bell Labs did not need to be overly concerned with advances outside AT&T, as they were the final arbiters. Many who have worked there described the extreme insularity, and, as one noted, "The world is Bell Labs. Everything outside was assumed to be behind the times, and they never looked outside to see it was falling behind."

Bell Labs never had an entrepreneurial bent; a focus on profits is not what drew many of the world's brightest minds to Murray Hill. Robert Lucky, a Bell Labs researcher and now vice president at Telcordia, remembers that "science was king, innovation was the Holy Grail, physics dominated, and nobody had much of [a] thought for business." Bell Labs' reputation can be summed up as "great technology, too late." For decades, the organization endured the well-founded criticism of being too slow to market and delivering overly expensive products laden with excess features and complexity. Bell Labs could not shake the reputation for having great minds, great research, and an inability to convert this quickly into marketable products.

This was not simply perception. It was AT&T that figured out how to make two windows operate on a computer screen simultaneously, but then it did nothing with it. It invented the cellular phone and then abandoned the business, only to buy it back later. Packet-switching technology to transmit data more effectively than traditional circuit switching was invented at AT&T but not developed further because the company feared it would cannibalize existing business. Historically, Bell Labs' track records for innovation and commercialization could not be more disparate.

Culturally, Bell Labs functioned like a university campus—from which most of its researchers originated—and commercial values were not highly esteemed. "Members of the technical staff," as they were known, wandered in and out of one another's offices in an informal university-like atmosphere. Eating, sleeping, and arriving at the office were not activities performed according to a typical corporate schedule. "There was no specific charter," Rajiv Laroia, an MIT mathematician who joined Bell Labs in 1992, explained. "We were given sort of free rein to do anything we wanted to do." Bell Labs was the ultimate ivory-tower institution where the world's most talented scientists and engineers were given carte blanche to pursue their intellectual interests in a well-funded setting free from the distractions of teaching, departmental politics, and the race for tenure.

The magic worked. The intellectual chemistry created by putting together great minds from every discipline in a single secluded building in a climate of unfettered discovery allowed ideas to flow freely. Bela Julesz, a Hungarian refugee who came to Bell Labs in the 1950s, described the unusual environment he found: "Here the idea is that distances are so constricted that you can, if you have an idea, rush over to someone who is an expert on, say, stochastic processes or invented I don't know what electronic device and consult him. It's so nearby that it takes you less time to go over and see him than to suppress the idea. In a university you would say to yourself, 'Shall I take my raincoat?' and by the time you made up your mind it's just as easy to suppress the idea. Also here, when you come over to a guy's office, he is pleased to see you. He regards helping you as part of his job and will even wipe off the blackboard for you. . . . I read something once that I still think is the best one-sentence description of Bell Laboratories. It is like a big baroque organ. If you are interested in playing one-finger accompaniments on a baroque organ, then you couldn't be at Bell. But if you have something to do where you have to pull out every register, then this is the place to do it."

As part of the 1984 divestiture, the 25,000 Bell Labs employees were divided between AT&T and the Baby Bells. The name Bell Labs stayed affixed to the doors of AT&T, but four thousand employees were moved into AT&T Information Systems and another three thousand became the research arm of the local phone companies. This research group was known first as Bellcore and later as Telcordia.

Arno Penzias, Nobel Prize recipient and vice president of research at Bell Labs from 1981 through 1998, remembers, "One of the things that people are constantly amazed at in this place is just how little short-term management there really is. . . . People don't believe that nobody tells me what to do on a daily, weekly or even a monthly basis. Occasionally I get a little bit of fatherly advice, but in the past year I do not think I have spent as much as four hours with my boss discussing how I should be doing my job. Instead we focus on our long-range objectives."

During the postdivestiture years, Penzias and Bill Brinkman, the head of Bell Labs fundamental research, anxiously watched the decline of Bell Labs. Although the competitive environment into which AT&T had been thrown had changed the company, Bell Labs had been largely unaffected. There was an ongoing disconnect between the scientific minds of Bell Labs and the marketing types at AT&T. Penzias felt that radical changes were needed and remarked, "AT&T senior management continued to invest more than one million dollars every working day in our work, and they give us the freedom to use it as we think best. That gives me a deep sense of responsibility to use our resources and our freedom wisely." Brinkman remembers that "what was very clear . . . was that we were in deep trouble. The business suits in AT&T were really coming at us saying, 'Hey, what in the hell are you doing for us?' The answer was we were just too researchy."

By 1989, Penzias was determined that Bell Labs needed to be remade into an organization that would support a competitive technology company. The wet sciences were out. Biology and chemistry were phased out over time, and physicists and mathematicians were told to find a way to make their work relevant to AT&T's wide range of businesses. The changes were not met with enthusiasm. A number of top-flight researchers streamed out of Bell Labs. Many of them found their way back into academia. Some felt that Penzias, a radio astronomer who had won the Nobel Prize in 1978 for detecting background radiation left over from the Big Bang, had betrayed the ideals of the labs.

When AT&T and Lucent parted ways in 1996, Bell Labs was again bisected. Again there was intense legal wrangling over which entities would continue to operate under the Bell Labs name and which would leave with AT&T's slimmed-down business. AT&T kept approximately 30 percent of the research capacity, renaming it AT&T Labs, and the remainder departed with Lucent under the Bell Labs name. Thus, the Bell Labs with Lucent's bright red innovation ring hanging over the door in 1996 was a very different entity from that which historians eulogize.

At the time of Lucent's birth, Schacht and McGinn made it abundantly clear that Bell Labs would be the heart and soul of Lucent. The

Lucent name did not go anywhere, not on a business card, a sign, or an advertisement, without its tag line, "Bell Labs Innovation," affixed. Schacht went as far as to relocate the new company to the longtime home of the Labs in Murray Hill. He observed, "We could have gone anywhere—downtown New York, out in the woods somewhere. This isn't the most convenient place to be; it hasn't been fixed in years, it's depressing. But we wanted to send a signal to the external world and internally that this was the heart of the place." As a both a practical and symbolic gesture, this allowed Lucent to integrate more closely its research capabilities into its operating businesses. McGinn held his first meeting as president of Lucent with the Bell Labs staff and described research as "the key to our survival and prosperity . . . the staff of life for us."

McGinn and Schacht believed that if they could remake Bell Labs, this would be their ace in the hole. To many other observers the situation was starker: if they could not transform Bell Labs, they simply would not succeed. During the road show, Lucent's management had wowed investors with the Labs' accomplishments and the possibilities that such talent held. Lucent had an early, easily understood, historically proven advantage; the message could not have been clearer. Other companies would have to seek out innovation, but Lucent already owned it. At every stop on the investor road show, management reemphasized its commitment to invest 11 percent of the corporation's total revenues on R&D, including 1 percent for pure research (although prestige accrued to Lucent in the opposite proportions). Success in the Labs would be entirely redefined; no longer would publishing highly acclaimed papers be a central feature on internal résumés. Success would now be defined by witnessing creations brought quickly and profitably to market.

Steven Levy, a bright, personable stock analyst from SolomonSmith-Barney and later Lehman Brothers, followed Lucent from its early days and soon emerged as the thorn in Lucent's side. Levy began his high-profile coverage of Lucent as an ardent supporter with an inside track on the new company after he had spent six years working in AT&T's Long Lines business. Levy had many friends at Lucent and understood the culture firsthand. In his earliest research he addressed Bell Labs' biggest problem head-on: "In the extreme, it could be said that the Bell Labs invented everything important in the communications industry but AT&T became the commercial market share leader in just about nothing." Levy labeled it the "Invention to Commercialization Dislocation" disease, and he hoped Lucent had recovered from it. From his vantage point as close observer, the necessary changes looked as if they were falling into place and investors had a concrete reason to be optimistic. He noted, "Lucent had transformed itself into a model of entrepreneurship. By that we mean contrary to its behavior of the past century. Lucent is now capitalizing on the Bell Labs 'product engine.' . . . Without these changes Lucent would remain just the owner of one of the most prolific research centers in the world and not the company that runs those inventions into profits."

Inside Bell Labs, researchers were nervous. AT&T had at least been the devil they knew, and it was no secret that Lucent's management had no experience running a large research organization. "When Lucent came about it scared the hell out of us," said Horst Stormer, a Nobel Prize–winning physicist. "We wondered, 'Are these guys rich enough to support us?' " For the Labs it was the third shake-up in a decade, and there was no reason to be certain at the outset that this one would be successful. "A lot of people were looking around for opportunities elsewhere and it seemed like there was going to be zero opportunity to do basic research of any kind," one researcher recalled. But the Bell Labs researchers need not have worried. Lucent tied Bell Labs' research budget to its revenue stream, and by 1999 the research budget had topped $4 billion, a far more generous allocation of resources, relative to the size of the business, than the Labs had seen under AT&T.

The pressure to change brought by management was intense, but Lucent was fighting the weight of Bell Labs' history. Lucent, as did AT&T before it, likes to brag that it has the only switch with the "five nines" standard: it works 99.999 percent of the time. Pick up the telephone receiver, and the dial tone is always there. The power is out; the dial tone is still there. The cable is out; the dial tone is still there. This was the standard to which Bell Labs and AT&T had always adhered. But this is not industry standard. Anyone who has ever operated a PC and watched in horror as it crashed in the middle of research knows that the commercial powerhouses of the late twentieth century operated with a much higher margin of allowable errors. Researchers who worked at Bell Labs under AT&T remember that there was no flexibility on the subject of quality, that it was unquestionably better to provide no service than to provide poor service. As one former researcher noted, "For Bell Labs, success is about 5 out of 5, about perfect accuracy, but it operates in a world that lives by a much lower standard, of perhaps 8 out of 10." McGinn was already a convert. He said, "We believe there's a lot more value in speed than in the notion of perfection, which is elusive at best." He thought that accelerating the pace at Lucent, but particularly at Bell Labs, was "actually very helpful because it forces people to the essence of the issue, the argument, the decision that you have to make, and also gets us away from the idea that it has to be perfect. Because when working at Net speed, what you're talking about is introducing the notion of human frailty." In the balance between speed and perfection, AT&T never had any doubts about where its concerns lay. For the newly formed Lucent, casting off this history, like shedding any deep childhood scars, would be an enormous and unending challenge.

Could the notions of speed and profit permeate Bell Labs after seven decades of being ignored? For some, the answer was "no." "I do what interests me and excites me," said Gerard Holzmann, a Bell Labs scientist. "I won't lose any sleep if nobody uses it. I don't work for Lucent; I work for Bell Labs. If this were to become a place like any other, you could work anywhere." While generations of researchers may have come to Bell Labs because it provided this setting, the more recent generations who have arrived at Lucent's Bell Labs got with the program and focused much more closely on the company's commercial needs. Bell Labs still employs researchers to work on projects with time horizons of a decade or more. This area of Bell Labs gets a great deal of publicity and prestige. And while pure research has been whittled down over the years, deep inside Murray Hill you can still find scientists conducting experiments with live slug brains to uncover basic principles for constructing biological computers or mapping the universe's dark matter in order to figure out how gravitational forces bend light. The challenge has always been to strike the uneasy balance between quality and speed, between pure research and application.

Every effort was made not to repeat the mistakes of the past, to give research the central role that the massive investment justified. But the key to success would be to capitalize on development opportunities as soon as they arose. "In the past, research has come up with something neat, but when we toss it over the fence to development people, they're not always there to pick it up," said Wayne Knox, director of advanced photonics research at Bell Labs. "Startups have always been good at picking up and running with research because they're focused on one thing. We're working to make sure that research makes it all the way through to development." In order to create a more focused environment, Bell Labs' senior management began to triage the work coming out of the labs. Every year, approximately thirty projects deemed to have the highest commercial potential (through cost reduction potential, improved functionality, or the creation of a new market) were labeled "breakthrough projects." These were given additional resources and staffing in an effort to increase their chances of success and cut the time to market in half. The process was highly successful. "Some of these have moved at a speed that has sort of dazzled us," said Mel Cohen, Bell Labs' vice president for research effectiveness. The Lambda router was just such a breakthrough project, and the first all-optical switch. Using 256 micromechanical mirrors, each the size of a pinhead, the Lambda router steers light waves from one optical fiber to another by slightly tilting the angle of each reflection. It moves traffic sixteen times faster, because it does not convert the signals back and forth to their electrical form. More than a technological leap, the product was proof that the triage process was working within Bell Labs and that innovative ideas would not get bogged down in a cramped warren of aging laboratories and small offices but would quickly see the light of day. Although the product was viable, it came too late in 1999. The market downturn after 2000 meant that in an all-out effort to cut costs across the company, many leading-edge technologies were canceled, and in the summer of 2002 the Lambda router project was shut down.

Another idea that made it out of Bell Labs in record time was the Ocelot. To date, cellular phone use bears more in common with the crash-prone PC than the good old-fashioned telephone. Anyone traveling in a car has had the frustrating experience of suddenly hearing nothing as the telephone conversation disappears in midsentence. For the

wireless carriers this has been a source of enormous customer dissatisfaction; for Bell Labs it was unacceptable. The problem is getting a
strong signal in all parts of the cell covered by the transmission tower.
To do this, engineers travel around the area testing and retesting the signal and then relaying the information to the transmission tower, where
another engineer can make the adjustments. This was a decidedly low-
tech methodology for solving a high-tech problem. As a more elegant
solution, Bell Labs invented Ocelot, a software program that could
mathematically model the geographic terrain of a wireless network and
allow a computer to determine the optimal number of antennas as well
as their direction and tilt. Ocelot could do in minutes what it was taking
technicians days and weeks to accomplish.

It was a testament to the interdisciplinary approach that has always
sat at the heart of Bell Labs. In a fortuitous lunchroom conversation, an
Ocelot researcher, Ken Clarkson, mentioned that the Ocelot prototype
was running too slowly and the changes he was hoping to make were
proving highly complex. Howard Trickey, a distinguished member of
the technical staff at Bell Labs' Computing Sciences Research Center,
overheard the difficulties and suggested a different and ultimately successful approach to the computer programming difficulties. After lunch
he joined the Ocelot team, and the product was ready for use in the field
less than a year later. Penzias noted the change in practice under Lucent,
saying, "They still create jewels, but more of them are made into jew-
elry."

Organizing and reorganizing Bell Labs in order to make it more re-
sponsive to the marketplace perhaps begs the question of whether the
expenditure involved in an enormous in-house research organization
should be made at all. Is such an institution anachronistic and unafford-
able at the speed at which innovation must now travel? There was a time
when such research labs focused on some of the largest scientific questions of the age. But companies had spent a decade retreating from research efforts whose financial contribution was often impossible to
discern. Shareholders no longer expected or even hoped that corporate
R&D labs would reveal the scientific truths of the age; they were simply
looking for a return on their investment. That Bell Labs was considered
a national treasure no longer seemed consequential.

Andy Grove, the legendary chairman of Intel, while remaking his company asked the question "What would you do if we got kicked out and the board brought in new management; how would actions be different?" It was a relevant question for Schacht and McGinn, who had committed to spending 11 percent of their revenues at Bell Labs, a substantial commitment but in line with industry norms of 12 to 14 percent for mature companies. They were bequeathed Bell Labs, and it is not clear that *de novo* one would establish such an enormous research organization attached to a high-tech conglomerate.

Publicly, Lucent's unwavering belief in the brainpower of Bell Labs bordered on the religious. "We had no problem with pouring over $3 billion into Bell Labs in the past year," said Bill O'Shea, executive vice president of corporate strategy and the somber scientist who runs Bell Labs. "We know we'll get a return on our investment. Bell Labs is the engine for this entire company." Lucent's management certainly conveyed this impression to Bell Labs. Yet privately some in senior management indicated that over time, if Lucent's business had continued on its growth path, spending on Bell Labs research would have been cut back, from the ongoing rate of 10 to 11 percent of revenues to 6 to 7 percent. In private correspondence, one senior Lucent executive argued that a large research lab was an unaffordable luxury that had to be whittled down. Cisco's successes with its "research by acquisition" model called into question the value of a large in-house operation. Perhaps more innovative product development might come from small, unfettered start-ups with nothing to lose and everything to prove. Some outside observers, and even many of those who saw Bell Labs from the inside when their companies were acquired by Lucent, argue that the Labs are living on their past reputation. Insiders disagree. "Every time something happens in the industry, as it is now, we get this resurgence of 'Bell Labs is dead,' " said Cherry Murray, senior vice president for physical science research. "We've always managed to come out of it stronger. Bell Labs is not dead. It is adaptable."

CHAPTER FIVE 1997

1997 Stock price   High: $17.38   Low: $8.57

> *A key to Lucent's success has been our willing-*
> *ness and our ability to constantly reinvent*
> *our company to succeed in an industry that is*
> *constantly reinventing itself.*
>
> Rich McGinn

In his earliest report covering Lucent, dated November 1997, analyst Steven Levy posed a challenge to his investor-readers:

QUESTION:

How does one turn a $20 billion company into an entrepreneurial entity?

ANSWER:

(a) Slowly

(b) As fast as one can

(c) Carefully

(d) With as much passion as one can muster

(e) By taking calculated risks on important people at all levels

(f) By removing it from its regulated parent

(g)   By doing it in a market environment that is growing so fast it forgives early mistakes

(h)   By underpromising and overdelivering

(i)   All of the above.

Levy quickly put his readers out of their misery by giving them the answer: "i." His bullish recommendations on Lucent reflected the fact that, from his close observation, Lucent managers were closely adhering to all his conditions. The right moves and good luck had taken hold.

Fortune could not have been kinder to Lucent. A confluence of events such as had never before existed combined to send off Lucent into some of the most favorable market conditions ever encountered by the telecommunications industry. Not since AT&T and its hundreds of early competitors had sought to bring phone service to a population that had never set eyes on the device had there been such optimistic projections for growth. While AT&T and Lucent management had voiced buoyant forecasts at the new company's inception, even they would be stunned. "Business just exploded," says Schacht. "It was a huge, huge favorable event that we did not foresee. Nobody saw it."

From the moment that Lucent left AT&T, its bottom line was enhanced by a myriad of factors that, like keys fitting precisely into their locks, would open the way for even more growth. At the time of the spin-off, Lucent was a quarter of the size of its parent. Nineteen months later, in April 1998, it had eclipsed AT&T's almost $100 billion in market capitalization. If Lucent had felt it had anything to prove to its progenitor, it had done it. Envy had seeped into the relationship, and as one Lucent senior manager recalled, "We were told that we were doing too well and that they 'wanted a piece of our stock price.' " AT&T asked for huge discounts on Lucent products and services and told Lucent's management, "You owe it to us."

It is impossible to overestimate the advantages created by a booming global economy. Economic growth alone would have brought Lucent success in its early years, but coupled with major technological advances and a flood of new capital investment, it created conditions in which the company exceeded all expectations. The calendar and the tide of history both added to this momentum. With the millennium approaching and the commentary replete with Y2K doomsday forecasts of a technologically induced meltdown, there was a surge in demand for upgrades in software and communications equipment. And some mileage still remained from the continued changeover from antiquated analog systems to new digital equipment.

The first wind of fortune to waft over Lucent was the massive expansion of the wireless phone business that took place in the 1990s. Cellular phones, in their bulky and expensive early incarnations, were the purview of high-ranking businesspeople. Few analysts predicted they would shrink to the size of toys and be used by teenagers to communicate across high school campuses. (Lucent's 1999 annual report shows an eleven-year-old bragging that his new mobile phone means he is never late for dinner.) The increase in the number of mobile phone users worldwide surpassed all predictions. In the decade between 1985 and 1995 the number of subscribers rose from less than a million to 90 million, and the best guesses at the time indicated that by 2001 there would be 330 million people attached to a mobile phone. Instead, by 2000, the actual number was already double that and the only question that remained—one that would have seemed ludicrous only a few years earlier—was the date when the number of mobile phones would surpass fixed landline phones.

Wireless communication may have been a new market, but to Lucent it was old hat. The new technology played into its greatest strength: designing and building equipment for the transmission and switching of voice communications and supplying component microelectronic parts for base stations and handsets. Grabbing and holding market share in the wireless business proved to be a huge success and one of the largest contributors to bottom-line growth. Lucent was stunned to find its wireless business growing at above 50 percent per year. Lucent had a huge financial and psychological win when Sprint, which had not done any business with AT&T, awarded it a contract for $1.8 billion to build 60 percent of Sprint's PCS (personal communications service) wireless network over a five-year period. The spin-off decision had felt like a good one from the start; the surge in Baby Bell business that began immediately vindicated the original reasoning. But in garnering business from the likes of Sprint, Lucent was able to give investors an early glimpse of the company it was to become.

Growth in Lucent's largest and most profitable business, the market for large circuit switches to handle traffic over traditional phone lines, had long ago been pronounced dead. Almost every household in America already had a telephone; where would new business come from? So the thinking ran in the early 1990s, when no one predicted the demand for second phone lines that would sweep the country. Suddenly, Internet service providers such as Prodigy, CompuServe, and AOL made a dedicated line for computer use a desirable consumer commodity, like air-conditioning in cars or microwave ovens. Sure, you could live without it, but who wanted to? The nature of Internet usage altered as it began to replace the encyclopedia, the evening news broadcast, and the book shop, making it easier to justify the expense. E-mail was fun, even useful; surfing the Web was entertaining, even mildly educational. Thus many families, even those without a teenager, discovered that the extra cost of a second phone line moved from the luxury to the necessity column in the family budget. Be it for the fax machine or the Internet, the lowly phone line had taken on a new life, and by 1996 there was an explosion in demand. And there was Lucent, the largest purveyor of the most reliable switching equipment ever built, refrigerator-sized 5ESS (electronic switching system) switches complete with the associated hardware and enabling software (call waiting, voice mail, caller ID, and other expensive custom services that delighted phone companies by generating a new stream of cash flow). The barriers to entry in this segment of the market, which had been built on generations of Bell Labs research, were a mile high. Quarter after quarter Lucent's management warned its investors that the switching end of its business, which contributed 15 to 20 percent of the company's revenues and vastly more of its profits, was a mature business and that investors should temper their expectations accordingly. Then, gleefully, management would report unexpected growth for the old workhorse switch. Although on a longer-term basis the circuit-switching business *was* in decline, the surge in second lines provided a temporary boom to a business with high margins and few competitors. It was not long before money from the rapid adoption of the Internet was dropping directly to Lucent's bottom line.

The switch itself held a vaunted position within Lucent even as other technology eclipsed its importance. Those who sold it were regularly feted, much to the chagrin of those who peddled Lucent's cutting-edge products. Dawn Truax, a longtime AT&T/Lucent salesperson, remembers the Achievers Club meeting when she learned the importance of the 5ESS. Lucent's most successful salespeople sat in a crowded ballroom listening to senior executives laud their performance. Those sales groups deserving special mention were called to the stage and given awards to commemorate the previous year's achievements. It was a global event, and Truax was seated next to a Sikh sales manager wearing a turban. As the speaker onstage enumerated the list of items the fifty-member sales force that serviced GTE (an independent phone company that later became part of Verizon) had sold that year, he gave special note to the eleven 5ESS switches (which would have sold for $5 to $7 million each). Truax was not impressed and mumbled under her breath, "That's less than one 5E a month." Her seat companion overheard her and, shaking his head at her naiveté, replied, "Ahhh, but you forget, those are 5Es."

Not everyone was impressed with the 5ESS business. John Dickson, CEO of Lucent's microelectronic business and then later the spin-off that became Avaya, warned McGinn that this business was not as good as it looked. He wrote, "The old switching business is also ripe for review. We have wasted years in not getting the proper (= higher) margins from the business. We have deluded ourselves that its high margins represent excellent performance as opposed to an organization that with drive and focus could achieve much more. Now is the time to really downsize aggressively and get its thousands of people working on our future as opposed to our past."

Americans were not the only ones suddenly to develop a taste for a second, or even a first, phone line. Global teledensity, the number of main telephone lines per one hundred inhabitants, was rising in most of the world, even as half the world had yet to make its first phone call. In some countries, demand for new telephone lines doubled and tripled. As if there were not enough good news for the newly created equipment maker, Europe deregulated its communications markets, opening the voice markets to full competition. In 1998, the first year after deregulation, five hundred new carriers were created; by the following year there were one thousand. Under government order in 1925, AT&T had retreated to the U.S. shores, so, Lucent's foothold in Europe was minimal. In unchanging circumstances, with competitors such as Siemens, Nokia, and Alcatel, it would have had a difficult time competing with the entrenched European equipment makers. But with the pie expanding so quickly on the heels of deregulation, Lucent found it much easier to gain market share. McGinn proudly announced in 1998, "We exited the international business in 1925. Since divestiture, we have built up our international business market share to 25 percent, exactly where it was 73 years ago."

Lucent's past, and its massive installed base of equipment, was a major key to its future. Even before the formal separation, the Baby Bells had begun to return to the fold, showering Lucent with new orders. In 1996, for the first time in a decade, Lucent was hiring in its manufacturing facilities because volume in its factories was up. In a market where new entrants sprang up almost every week and others disappeared with alarming frequency, durability became an increasingly valuable asset. Lucent had the financial strength and the proven ability to execute. This gave customers making multimillion-dollar purchases the confidence they needed to sign on the dotted line. Lucent would be there, its service people would come when they were called, and, given the historically low turnover among AT&T/Lucent employees, as often as not they would be the same people. Although start-ups might dazzle the marketplace with great technological leaps, service providers spending hundreds of millions of dollars on equipment that could last a decade wanted to purchase it from companies that they knew had the experience to design and construct an entire network, not simply ship the black boxes. Advantage Lucent.

From the start, Lucent's businesses had too much overhead for their anticipated cash flow. Even before the red logo was hung over the lobby, Lucent would need to face some deep budget cuts. AT&T had been a multitiered organization with a bloated cost structure. Lucent's cost structure was out of line with the industry's, stemming from years of operating in businesses in which it dominated the market and cost was not a primary concern. "Sales, general and administrative," the catchall line item for cost, showed that Lucent's expenses were running at 25 percent, a staggering rate compared to Nortel's 17 percent. Each manager was asked to provide plans for a dramatic downsizing, as the new company would cut a sixth of its workforce. "We were taking half of

AT&T's people but only a quarter of the revenue: we had to figure out how not to scare these people half to death," Schacht recalled.

Schacht and McGinn promised shareholders that they would reduce their costs and improve their margins by a massive 4 to 8 percent over a three- to five-year period, although neither of the company's leaders would be pinned down in terms of timing. They began to hack away at corporate excesses, put the company on a spending diet, downsized unprofitable departments, slashed the cost of information systems left over from the AT&T days, outsourced some of this work to IBM (and sent the staff along to do it), consolidated operations into fewer locations (saving 2.5 million square feet of office space), and reduced its tax bill. The effect on income of such a dramatic decline in overhead began to show up immediately. These quick boosts to Lucent's profitability could not be replicated, but they helped get the company off to a fast start.

To some the constant talk of cost cutting might have been seen as simply good publicity, but McGinn was serious. In the separation from AT&T, McGinn had asked each manager to submit a timetable for the orderly distribution of inventory in warehouses that were being closed or sold off. He expected his managers to take this responsibility seriously and assumed he would receive quick responses. But as McGinn surveyed reports of stagnant inventory piled high in obscure warehouses all over the country, he was thoroughly frustrated. Did his managers not understand how cost conscious the new company needed to be? The new president wanted accountability; he would demand it; and he was going to Indiana to drive the point home. Fed up with demanding action and seeing no results and still in possession of AT&T's jets, McGinn decided to take his executives on a little field trip. The executives knew they were not being transported to La Porte, Indiana for good behavior. When they arrived, they climbed into an AT&T van awaiting them; the mood was tense. At the huge warehouse, the six chastised managers followed McGinn from pallet to pallet, eyeing the heaping piles of equipment. There were mountains of unaccounted-for inventory for which no one in New Jersey wanted to take responsibility. With clipboard in hand, McGinn pointed out to each of the managers which inventory was theirs and the fact that it was sitting there with no plans for disposition. Some of the stuff was filthy, covered in many layers of grime. McGinn was not done with his little lesson. He climbed behind the controls of a forklift and moved the pallets toward the loading docks as someone snapped his picture. Later he was presented with a framed copy of the photo with the caption "Fermez la Porte."

Lucent's profits were also bolstered by reductions in the overall head count. Only months before the spin-off, AT&T had stepped into a major public relations morass when its board announced enhanced compensation for Allen at the same time as the layoff of forty thousand employees. Few were impressed with this exchange, and a storm of bad publicity rained down upon the long-distance company. Lucent took this lesson to heart and used caution in its staff reductions. Schacht was proud of protecting his employees: "I think we reduced about 24,000 people. Of that there were less than a thousand who actually lost jobs. The rest of them were sold in divisions. There were temps that went off to different work. There were attritions, we didn't hire people. There were some layoffs but the minimum; there wasn't any great big fuss. We invested in the company. The stock price has tripled. And I just defy anybody to say we would have done better by slash and burn, no damn way."

For investors, who were themselves buying cell phones, logging on, and ordering more phone lines, the Lucent story was easy to understand: talented and experienced management already delivering on its promises, an obvious increase in demand for well-understood products. By 1998, Lucent looked all but unstoppable, dominating the top spot in worldwide telecom equipment sales. McGinn was unabashedly ecstatic about his company's prospects, saying, "It may be a result of good planning or just serendipity, but it's absolutely phenomenal to be in this business right now."

CUSTOMERS?

In the wake of the bankruptcies and scandals that have rocked the telecommunications industry, it is hard to remember that it was once a steady, moderate-growth industry interesting only to those it employed. Before 1996, the one remarkable event in the industry was the breakup of AT&T, and then only because it was a leap into the void. In 1996, Congress pushed the industry into the void again with the first major piece of legislation to tackle telecommunications in sixty-two years, legislation that intentionally set off a free-for-all, forever blowing apart the vestiges of a regulated industry. While AT&T had faced a throng of competition in the long-distance market since the early 1980s, local telephone services was still virtually a regulated monopoly. In 1996, the Federal Communications Commission decreed that competition should be completely free and open in the local, long-distance, and wireless markets and that market forces should prevail. The goal was to carve out a level playing field for new and entrenched service providers. The government believed that the outcome of these changes would be lower rates for consumers and quicker deployment of technologically advanced services. It speculated that newcomers, as well as the large long-distance service providers and cable operators, would enter the local phone market, driving prices down and improving services. And it was correct; all this happened, and more.

The Baby Bells were unprepared. They had watched AT&T endure competition, grateful that it had not been their fate. But in 1996 Congress gave a leg up to a new breed of competitors, the Competitive Local Exchange Carrier (CLEC, or "see-leck"), and required the existing local phone carriers to help these arrivistes get started. CLECs, financed by the capital markets, set up to compete with the incumbent phone carriers by providing their own networks and switching at reduced cost. To do this, the industry needed to be restructured. The former model, whereby a service provider built a network with astronomical fixed costs and then imposed high usage fees in the form of telephone bills to recoup these costs, needed to be altered for competition. Without millions of users, it would be impossible for this model to work. The Telecommunications Act of 1996 reversed this law of economics by legislating access to the existing networks. The Baby Bells were required to open their networks to the CLECs and to negotiate interconnection agreements in good faith and on "just, reasonable, and nondiscriminatory terms and conditions." Baby Bells were forced to sell their services at wholesale rates for resale by the CLECs at retail prices. While at first glance this looked like a blow to the incumbents, who had paid for this network and were simply trying to recoup its cost, ultimately the CLECs would be at the mercy of the Baby Bells.

Early entrants into the communications industry, including Sprint,

MCI, Teleport Communications Group (TCG), and MFS Communications, had been fairy-tale investments with happily-ever-after endings for their investors. In each case the new company creamed off some of the best corporate customers, provided state-of-the-art technology, and faced minimal competition other than the Baby Bell in their region. MFS established a fiber-optic network through forty-five U.S. cities that allowed long-distance providers to bypass the Baby Bell networks, a development that proved irresistible to WorldCom. WorldCom paid $14.4 billion for MFS so it would no longer need to pay the "Baby Bells" for the local connection, despite the fact that MFS had only $600 million in revenue in 1995. WorldCom and its bankers valued MFS at more than six times the value of the fiber MFS had buried. It was an unbelievable investment return for those who had put their money into MFS early, and it lit a fire under an entire industry. Within a decade, telecommunications had morphed from a regulated utility to profit opportunity bar none. TCG was founded in 1982, and by 1997, when it had laid ten thousand miles of fiber-optic cable connecting eighty-five cities, it appeared on AT&T's radar. AT&T happily forked over $11 billion for TCG, even though the latter had less than $500 million in revenues. A benchmark for the industry was firmly established. The model was now in place: make a substantial capital investment almost entirely through borrowed funds, and in time a larger, established telecom company will pay an enormous multiple of the original investment. The older companies had customers, revenues, and networks that were sometimes thirty to forty years old. The newcomers had cutting-edge technology and an appetite for risk and debt. It looked like a match made in heaven.

The early success stories formed the basis of much of the deluge of investment that eventually buried the telecom industry. Although the reasons for the early CLEC successes were not entirely comparable (for example, MFS and TCG had only the Baby Bells to worry about, not hundreds of little MFSs and TCGs running around in their market), the stock market poured billions of dollars into telecom IPOs between 1996 and 2000. In 1996, $6.6 billion of public capital was raised for emerging carriers; the following year this rose to $8.2 billion, and by 1999 it had exploded to $82.2 billion.

The CLEC business model appeared sensible, at first. With reasonably priced access to the existing local networks, the ability to cherry-pick only the most lucrative accounts, and the adoption of advanced technologies for data transport, the CLEC model was compelling. Early successes reinforced this thinking. The market looked hospitable to the first CLECs, but, after only a year or two, they found themselves facing other CLECs, not simply the Baby Bells, in a battle for customer business. As industry analysts Bart Stuck and Michael Weingarten noted, "As a result [of these powerful advantages] being an early adopter CLEC fighting against a sluggish incumbent hamstrung with lots of regulatory constraints was a great way to deceive yourself into thinking that you were smarter than the rest of the world."

Thinly capitalized and highly ambitious, the new players were unburdened by the sunk costs of a large installed base and unbounded by any preconceptions surrounding the telecommunications equipment market. They were simply looking to establish new and profitable relationships with vendors who would sell them equipment and sometimes lend them the money to buy it. These new entrants, backed by venture capital, junk bonds, and the proceeds of their IPOs, went on spending sprees. The faster they built their networks, the faster they could start stealing Baby Bell customers to pay back the debt incurred in building these networks. Investment by new and existing carriers rose at the rate of 30 percent per year, or far more than double its historical trend. Capital spending on investment in telecommunications topped $350 billion between 1996 and 2001 and was nothing less than a godsend to Lucent.

But the uncertainty in this market meant that Lucent needed to answer the nontrivial question of who its customers would be. Would the Baby Bells end up as dinosaurs, their former networks picked over by thousands of CLECs, or would their incumbent advantage continue to serve them and allow them to beat back the challengers? Converging networks that could carry voice, data, and video were an accepted truism, but who would build this national broadband network and how it would be built was far from obvious. For Lucent, the answer to this question meant everything. Lucent dominated sales to the vestiges of AT&T, but as the 1990s wound down, this was the slower-growing sector of the market. The CLECs had different equipment needs. To win this business, Lucent needed to design and build equipment that suited

them. Lucent's strategy, the allocation of its human and material resources, depended upon correctly guessing the outcome of this battle for
dominance.

Plan A would be simple. Lucent would assume that the Baby Bells,
with their huge installed networks and millions of satisfied paying customers, would prevail. The spin-off would be a homecoming of sorts.
With AT&T out of the way, the two old friends could openly embrace
each other as business partners. There would be no muddy cultural waters to navigate, as everyone had Bell DNA and spoke the same language.

Bell Labs was geared up for the Baby Bells' networks. Its research
and product development was aimed almost exclusively at the needs of
large regional carriers. Lucent's product and sales teams were familiar
with the inner workings of the carriers, and, despite the tensions of the
previous eleven years, this was a business they understood. It was in-
comparably easier for Lucent to anticipate the equipment needs of in-
stallations that were largely theirs in the first place.

Maybe the local phone companies, despite little historical precedent
for locating the cutting edge, had grown up since divesture and would
use the full extent of their resources to face the upstart threat. Cash flow
would be king, and while the new kids might scrounge a few million
here and there from venture capitalists looking to get rich quick, nothing
could match the power of tens of millions of customers who paid their
phone bills every month. The local phone companies might be dinosaurs, but as the head of one Baby Bell pointed out, dinosaurs were
around for a very long time.

Plan B would be far more complex and would stretch Lucent's resources and expertise almost to their limit. The Baby Bells, or "Bell
heads" as their competitors liked to call them, would move too slowly to
fend off the more agile CLECs. They would underestimate the competitive threat, and then, when they finally grasped it, it would be too late.
Starting from scratch, the new players would be able to lay down brand-
new networks with all of the latest capabilities while the Baby Bells sat
and pondered the costs of backward integration into their existing systems. Those new to the game would have lower cost structures, the
thinking ran, and thus be able to outcompete on price. The growth rate

of the CLECs would be many times greater than that of the Baby Bells, and, while they started from a much smaller base, in time they might come to dominate the market. History would repeat itself, and the Baby Bells' market share would steadily decline, following the example set by AT&T. Those equipment suppliers hoping to develop long-term relationships with the future market leaders would have to jump in early and aggressively to compete for business.

Under this plan, Lucent would enter a brave new world where the rules and the technologies were far less certain. The company would need to forge new relationships in an environment where it had few contacts and little expertise. With a vastly expanded sales force, Lucent would try to garner business from the thousands of new companies all over the world that had begun to offer voice and data transmission in local areas. Immediately, it would need to repackage its existing equipment for these smaller companies with more modest and specialized needs. Lucent had not embraced the technologies, such as packet switching, that the CLECs needed to the same extent as some of its competitors; as a result of the influence of AT&T it had ventured down different routes. Bell Labs had the technology to commercialize packet switching, but AT&T had not had the will. Sabotaging its premier equipment product was not on AT&T's agenda. Mel Cohen, who recently retired as Lucent's vice president of research and effectiveness after thirty-four years, explained that "AT&T had a well thought out reason for perpetuating circuit-switch telephony, so they wouldn't necessarily be enthusiastic about driving the market for packet switch telephony." Bell Labs scientists remember that Network Systems held back on the development of faster optical switches because much of the fiber AT&T had in the ground would not sustain it. This left Lucent ill prepared to service the fastest-growing sector of the market. Lucent's product development would need to be retailored to focus on companies in the throes of their initial development. Schacht estimated that it would take two years to retool Bell Labs to enable it to develop the products CLECs needed. In the meantime, Lucent would need to acquire companies that had some of these products available immediately.

If the CLECs succeeded, after some inevitable industry shakeout, the future looked promising. Their aggressive approach might transform the industry, and their endless thirst for new technologies could provide the buoyant growth the stock market was coming to expect from Lucent. The investments were enormous, the anticipated growth path just short of vertical, and the spending had already begun. Like the Internet start-ups and electronic retailers, CLECs were a huge gamble, an industry with no track record, untested business plans, and weak finances. While many of them were run by managers with strong telecom backgrounds, others set sail with untested pilots.

The CLECs' thirst for cash would be unending. It was understood that the infrastructure build-out would take many years and they would operate for an extended period of time as loss-making entities dependent on the market for low-grade debt and the kindness of equipment vendors. For equipment manufacturers, this would mean longer-term accounts receivable and the need to provide billions in vendor financing. The credit quality of Lucent's newest customers would be lower than that of its traditional customers. Lucent's own success would rest on its ability to select the winners and losers in a nascent industry, because those who failed to thrive would leave a trail of bad debt in their wake.

Lucent's management made a bold bet on Plan B, that the CLECs would survive as the customers of choice, and began to adapt the company to suit their needs. It was a risky strategy but one that began to pay off from the start. In true non-AT&T form, Lucent took an aggressive approach to earning CLEC business. Senior management was so thrilled with the outcome of this aggressive sales strategy that one day they would literally kiss the feet of one of the sales executives who had executed it.

The Telecommunications Act had thrown the industry into disarray, with unending controversy about which technologies and companies would prevail. As Schacht had predicted, a dense fog had enveloped the landscape, and, although the CLECs looked to be taking the early lead in a business that knew no bounds, the critical question as the mist of the Telecommunications Act of 1996 lifted would be: Was it real, and, most crucially, would it last?

Lucent management must be lauded for an intense effort to reshape the company's culture, but ultimately it may have been hampered by its own good fortune. In 1997, Lucent announced that annual profits had increased by 44.7 percent to $369 million, or more than double analysts' expectations. Analysts love this sort of surprise, and it turned their growing affection for Lucent into true passion. As the adulation grew deafening, it reduced the imperative for change. Good times mask a multitude of sins, and many weaknesses would rear their ugly heads only in times of stress. The string of successes made it difficult for both those inside and outside Lucent to discern whether the company had truly created the entrepreneurial culture and environment it would need to compete in difficult market conditions. There can be no doubt that Lucent emerged from the spin-off reenergized and aggressive, ready for competition. But had top management done enough?

CHAPTER SIX   1998

1998 Stock price   High: $43.61   Low: $14.06

Since 1995, we've taken a $20 billion business and, through 1999, moved it to almost $40 billion, and taken profits from $500 million to almost $4 billion. We describe that as directionally correct.

Rich McGinn

Later, Rich McGinn would endure almost universal criticism. Shareholders, banded together in a class action suit, would argue that he had been misleading. Former employees would claim he had ruined the company, destroying their life savings in the process. Schacht, his predecessor and successor, would say that he had tried to run the company too fast, rev the engine too hot. But all that would be in hindsight.

By 1998, Rich McGinn was already a corporate superstar. *Worth* magazine ranked him eleventh among the best CEOs in America, a member of what they called the "new celebrity class." *Time* described him as a "smart, quick-thinking former history major who has turned Lucent into a telecommunications heavyweight and one of the biggest success stories of the 1990s."

The chorus of those singing Lucent's praises had almost no dissenters. *Fortune* said, "Lucent resembled an unblemished portrait that was hidden in the attic." Steve Levy, the Lehman Brothers analyst who would become an early critic, was still positive, speculating that Lucent would capitalize on the Bell Labs product engine: "In fact we believe that Lucent has put in place organizational, cultural and perhaps most importantly compensatory changes to its corporate structure that should enable it to be a leading edge commercializer of technology, not just a leading edge inventor."

Although only two years old, Lucent was being compared to the best technology companies of all times. As one industry commentator noted, "Just as IBM was to the '70s, Microsoft to the '80s and Cisco to the '90s, Lucent is positioned to be the technology company that defines a decade—and quite dramatically, the beginning of a millennium."

LUCENT'S METAMORPHOSIS

On January 20, 1998, only three months after passing on the title of CEO and a few months ahead of schedule, Henry Schacht relinquished the chairman's title. The transition phase was over. McGinn's hold on Lucent was finally complete, and his ability to remake the company was for the first time unencumbered. McGinn asked Schacht to remain on as a member of the board. While he hoped to expand the size of the board and alter its composition (increasing its relevance to Lucent and its industry), he would ultimately add only a single member. Lucent was short of warm bodies to fill the directorships, and McGinn had sometimes looked to his predecessors for advice. Viewing Schacht as one of his staunchest supporters, he asked him to retain his seat. This was a decision that McGinn would come to regret.

As the world in which Lucent competed changed radically, McGinn was determined that the transformation in his company would be no less dramatic. The telecom equipment business had once been a somewhat competitive manufacturing business globally in which Lucent (as Network Systems) and Nortel were effectively duopolists in the North American market. Both were large, lumbering companies with high public profiles and similar customers, and Lucent was rarely surprised or completely outflanked. The lag between research and product development was so long that few surprises were possible. Nothing in its decade and a half of life facing Nortel could prepare Lucent for the world that it entered in the late 1990s. Nortel had reinvented itself, becoming a more aggressive and innovative competitor. No longer was the North American market comfortably divided between these two giants, with the occasional incursion by a European or Japanese rival. Small groups of engineers, some of them Bell Labs alumni, armed with stores of free-flowing venture capital, temporary offices, and fewer than one hundred employees, were now at the forefront of technology. It was just the type of fertile intellectual environment that capitalism, in its best and healthiest incarnation, can provide and that can be witnessed at the outset of major disruptive technological breakthroughs. While no one would suddenly appear on the scene to build a new 5ESS switch, competition in the fastest-growing markets at technology's edge could and would come from anywhere.

All around him McGinn saw small, fast-moving businesses streaking through the design and development stages and bringing products to market while Lucent patiently listened to its customers reel off a shopping list of specifications they wanted from any new products. These tiny, nimble rivals, unfettered by a hulking organization, were picking off market share, particularly in the CLEC market, which was itself on a lightning-fast growth path. The increased competition looked certain to intensify and accelerate. Even if the engineers who were developing products were not in enough of a hurry to get those products to market, the venture capitalists bankrolling them urged them on. To McGinn this was the new face of competition, the revved-up world in which Lucent would have to compete. Lucent would have to either take on these new competitors or buy them.

For two years, along with Schacht, McGinn had exhorted his troops to pick up the pace, but words could do only so much. Although the executive duo had labored to change their company's worldview and create a more dynamic culture, McGinn knew they had not moved far enough. The 1997 employee survey represented not evidence of a strategy that needed to be refined, but rather a clarion call to action. Despite jibes from journalists, McGinn liked to say, "We're reinventing ourselves as a startup. The fact that we have lots of zeros at the end of our income statement is incidental."

McGinn was done exhorting, done with gestures and newsletters, with internal broadcasts and endless meetings. He could not mandate change, but he could reorganize Lucent in such a way that those areas dragging behind would no longer have anywhere to hide. Without dramatic change he would not get dramatic results. McGinn broke Lucent apart into what he loved to call "eleven hot little businesses." Critics inside the company said he acted too quickly, that the changes were too radical; but if he waited for a consensus, carefully getting everyone to sign off, it would be business as usual. The eleven new units were:

1. Microelectronics
2. Business Communications Systems
3. Data Networking Systems
4. Global Service Provider Business Sales
5. Wireless Networks
6. Switching and Access Systems
7. Optical Networking
8. Network Products
9. Communications Software
10. Intellectual Property
11. New Ventures Group

Start-ups have transparency. Without layers of organization or overlapping responsibilities, there is simply no place to hide. Managers cannot deflect the responsibility for less-than-stellar performance. The eleven small businesses would give each manager ownership and force individual responsibility on a company long used to making decisions by committee. McGinn's plan was to strip back the bureaucracy and force operating units out of the safe arms of a mammoth company into closer contact with customers and competitors. As a forty-nine-year-old CEO, McGinn found his youth blocked the road to the top for his young management team. The new structure allowed him to distribute internal CEO titles to those to whom many external opportunities beckoned. McGinn was looking for speed and accountability, and he told his management that he hoped his plan would encourage greater risk-taking and higher aspirations. If he succeeded, this reorganization would create an atmosphere more conducive to incorporating the flood of entrepreneurs who would walk through the doors as Lucent acquired companies over the following two years. Most important, he believed it would stimulate growth. If he failed, the new structure would create chaos and confusion.

The next best thing to becoming a start-up was to buy one. Once upon a time well-seasoned CEOs made acquisitions only occasionally, and only after careful study of the target, its products, culture, and aspirations. That had changed. Lucent devoured thirty-eight companies in its first five years, while Cisco managed to scoop up an incredible seventy companies in eight years. It was considered a truism at the time that a buying binge was necessary to the success of a high-tech company. Too many major innovations originated from small and medium-sized start-ups for the CEOs of larger companies to believe in their own self-sufficiency. Buying fast-growing companies with innovative products instantly increased market penetration and capitalization. Unless a company actively acquired small cutting-edge organizations, their intellectual resources would flow to its competitors unimpeded. Thus the race was on.

Scooping up thirty-eight smaller companies in the space of only five years does not come without a price. Senior Lucent executive Janet Davidson described Lucent's growth strategy this way: "We were literally asking ourselves to grow a mid-sized company every year. The things that drove the organization [were that] we didn't have to focus, because to grow that much, we have to serve everybody. You can't afford the luxury of focus. To serve everybody, you have to invest in everything, and we did."

In the battle for acquisitions, McGinn was at a disadvantage during Lucent's first two years as the U.S. tax code tied his hands behind his back. During this period Lucent, as a spin-off, could not use the "pooling of interest" accounting method to report its acquisitions. Under this method, the acquired company is taken onto the books of the buyer at its book value, or historical cost, rather than the actual cost paid. This allows the acquiring company to hide the premium it paid or overpaid for the company by pretending that the cost was far less. The acquisition appears to be a success from the start because pooling of interest reduces the amount of assets the acquirer needs to depreciate. The Federal Accounting Standards Board has since eliminated this practice.

Walking into a pizza parlor in 1998 and overhearing staff shout back and forth about the stock price of Lucent that day was highly gratifying for a CEO who only two years earlier had been running a division buried deep within AT&T. Schacht and McGinn both delighted in at the steep run-up in Lucent's stock price. When announcing the first stock split in February 1998, McGinn happily reminded investors that the stock price had tripled in less than two years. Even before leaving AT&T, Schacht and McGinn had been fully aware of Lucent's deficits in technology. Now the stock market had given them the means to remedy the situation. Acquisitions were completed not only for the technological and human assets they garnered, but also for the financial advantages they brought. Buying smaller, faster-growing companies with growth prospects far in excess of anything the larger company could or would achieve allowed CEOs to step onto an Escher-like staircase that led only upward, a virtuous cycle that in rising markets appeared to have no end.

Through a process that can only be described as financial alchemy, McGinn and other acquisitive CEOs were able to boost their share prices by purchasing smaller companies with higher P/E multiples. When a CEO of a company with a large sales and distribution capability bought a small company with a promising technology, the market assumed that the sales of that product would skyrocket. This caused the stock price of the larger company to rise sharply. Companies such as Lucent, Nortel, and Cisco went on a buying binge in the late 1990s, and Wall Street and investors all over America gave them standing ovations. Armed with this higher stock price, the newly enriched McGinn and his counterparts could repeat the process, buying ever-larger companies with ever-more-inflated P/E multiples, some with real revenues and customers, others with not much more than a promising idea, and find themselves further rewarded by the investment community. The high-P/E environment that prevailed during this period meant that even if the engineering did not work, the math did. This mind-bending equation made it much easier for Lucent and others to contemplate acquisitions that in an earlier, more cautious era, one in which speed was not the watchword, might not have been given a second thought. And so it ran, with stock prices escalating, a bidding war for fast-growing start-ups heating up, and CEOs rewarded almost equally for their poor investments as for their insightful ones. In this context there was little to be gained from the intensive evaluation of prospects that once had taken place. Management could easily become deluded about its own ability to judge such investments or even to build a business when the market rewarded them at every turn.

It was difficult to anticipate what Lucent would do, but in 1998 McGinn did not sound like a CEO who planned to shop until he dropped when he explained, "We've said that in the main, we're going to grow organically because we don't think we're limited by opportunity but rather by execution and the speed issue. . . . We're not opposed to acquisitions as long as they make good sense within the context of the business." Bill O'Shea, in his role as executive vice president of corporate strategy and business development, implied that Lucent would not need to go on an acquisition spree in order to find a source of innovation: "Everyone else had been content to rely on M&A to acquire technology, but no one has the kind of research organization we have." This was not an insignificant point. But even as Lucent spent between 10 and 12 percent of its revenues on in-house research, despite O'Shea's cautious comments, it also kept up with the Joneses and went on a buying binge for new companies.

During Lucent's first two years, the company laid out cash for a few modest-sized acquisitions. These early deals were a series of small, successful bolt-on acquisitions that broadened the product portfolio. Some of the companies were embryonic start-ups that Lucent knew well. As McGinn explained, "There are a lot of new companies coming up that are not real companies. They're really R&D efforts that need the value of distribution that we have, the customer relationships that we have, and are happy to get magnetized to us." These were some of Lucent's most successful acquisitions, small companies with fully developed products, employing 250 people or less, that could make a quantum leap in sales by integrating their offering into the Lucent lineup.

Yurie Systems, a maker of asynchronous transfer mode (ATM) access equipment, which Lucent bought in mid-1998, was just this type of company. Founded by Jeong Kim in 1992 and named after his eldest daughter, Yurie grabbed the number-one spot on *BusinessWeek*'s 1997 list of the "Hot Growth Companies in America." If Yurie had not yet appeared on Lucent's radar screen, it could hardly be missed once a picture of Kim, smiling and single-handedly hoisting over his head the prototypical communications black box, graced the magazine's front cover. The company had recently completed an initial public offering, free from debt and venture capital funding, and recorded an astounding growth in sales of 385 percent and earnings growth of more than 400 percent. Yet Lucent's success with Yurie had far more to do with Kim's own business experiences and the culture he had developed at his company than with any piece of equipment it sold.

Kim emigrated from South Korea to the United States at the age of fourteen and found the transition anything but easy. "We came here to pursue the American dream, but it's hard to keep your focus. People make fun of you, because of the language barrier and the fact that you look different. When you're a teenager, you're emotionally affected by these things. I would come home with a nosebleed, from the stress," he remembered. The challenging experience of entering an American high school without speaking English, living in subsidized housing, sometimes without enough food, and later being kicked out of the family home as a teenager gave Kim resolve for life. "Once you have some food and clothes to wear, everything's extra," he said. This rough start only fueled his ambition. During high school in Maryland he worked from 11 P.M. to 7 A.M. at a convenience store while living in the basement of the house of his math teacher, who had taken him under his wing. Later Kim earned undergraduate and graduate degrees with honors at Johns Hopkins University, while at the same time he was a full-time partner and design engineer in a computer firm. After college he joined the U.S. Navy, because, in his words, "I wanted to pay back this country for the opportunity it has given me." Kim became an officer in the prestigious nuclear submarine corps, and it was the Gulf War that gave him the inspiration for his company. "If we had a communications network that allowed us to communicate to the bomber in real time when the target has moved, the pilot could retarget," the soft-spoken Kim explained. "He wouldn't waste his trip." He plowed his life savings into a computer company that later failed. "All my paycheck went into the company just to keep it running," he recalled. "That was a mistake. We made a lot of mistakes in that company. Maybe it was a good thing, because I learned a few lessons about cash flow and so on."

In 1991, two years out of the navy, he earned a doctorate in reliability engineering at the University of Maryland, but starting a new business had become very difficult. He said, "It became very obvious to me that the communications revolution was going to be even bigger than the computer revolution. The world will have to go global in terms of communicating and connecting. I realized that was a very significant event, so I tried to team up with a couple of people to raise money and start a business in this field again. I couldn't do it. I was unsuccessful because, through the eyes of venture capitalists, I was a guy who came out of the military, with no track record of running a business. They don't care what you did in college. That was a long time ago. I didn't have any money. Most of my savings had gone into my old business." Kim set himself up as a one-man shop, hoping he might attract attention and eventually some funding. Yet several times the company struggled, running low on cash. Each time Kim took out personal debt to keep his business afloat. "I said to myself I would do this thing for two years. If I don't get headway, I'll rethink," Kim said. "I was determined to go at it for two years."

While many entrepreneurs sold their companies to Lucent and almost immediately made a U-turn and rejoined the start-up world, Kim stuck it out. He told McGinn from the start that he was looking for a new challenge. Despite the fact that the $1 billion all-cash deal with Lucent had earned Kim the twentieth spot in *Fortune* magazine's "richest 40 under 40," he hoped to succeed in the ranks of a large multinational corporation. Kim knew of AT&T/Lucent's troubles with commercializing the developments that came out of Bell Labs. At Yurie he had successfully been through this process, and he thought he might bring his new experience to bear. "They offered me a job as President of Carrier Networks," Kim said at the time. "Typically, when a big company acquires a small one, the top guy at the small company gets to run that part of the organization or leave. Lucent gave me two other major divisions. I'm responsible for the entire product line for the carrier networks, so I can be much more effective, I can add a lot more value. I can't tell you for how long, but until my creative juice runs out. That's where I think I can make the most contribution."

Kim's commitment to his new employer trickled down through the ranks of Yurie. From the start Yurie's president, Harry Carr, an AT&T alumnus, let his staff know that fence-sitters were not welcome. Their indecision was debilitating. He told his staff that he and Kim were staying and that they intended to lead by example. Everyone else should make up their minds, now, and he would wish them well if they left, but after that he did not want to hear about any more resignations.

The Yurie acquisition was executed correctly, and the timing was ideal. McGinn and Carly Fiorina met with the entire company and welcomed them to Lucent. At a company wide meeting shortly after the acquisition closed, they highlighted the acquired company's capabilities for the larger organization. With a market capitalization of just over $1 billion and only 238 employees, Yurie was at the perfect stage in its development to be picked off by Lucent.

There is perhaps an ideal moment to make a start-up acquisition, the exact point in time when it passes through concept prototype and the first customers show interest but before a corporate structure, marketing, and sales organization and a lengthy list of clients are in place. No customers, no revenues, and no products makes the company a shot in the dark, but a company in possession of any of these things comes with a fleet of executives and venture capitalists, all looking for their slice of the stock's multiple. Without the trappings of a full-fledged company, integration into the larger entity is easier, bringing fewer job redundancies and less dislocation. Acquisitions can look premature right up until the moment that they are too late and a competitor steps in front of you in the checkout line. There is an ideal moment to acquire, but it is excruciatingly difficult to pinpoint.

As a wall of investment capital money washed over the telecom sector, start-ups were cajoled into selling themselves earlier and earlier by investors impatient for an exit strategy. Venture capitalists had traditionally invested in companies with the hope of eventually making their way to the IPO market. Now they were cobbling together research efforts that were never intended to be stand-alone companies but rather were built to be sold to larger companies. These nascent start-ups posed a far greater risk for acquirers, because they had short track records and barely developed products. A number of Lucent's least successful acquisitions fell into this category. With products still in the development and testing stages, they required more R&D resources and lost focus and momentum within the larger organization. Cisco CEO John Chambers pointed out that the competitive environment for buying start-ups was an unfortunate turn of events, saying, "We bet on products 12 to 18 months out. We took dramatically higher risks."

Most entrepreneurs sold their companies to Lucent in part because they and their financial backers were looking to cash out, but also because Lucent could provide them with the infrastructure and distribution channels they needed to vastly increase the sales of their product. Lucent had the long-standing client relationships and installed network that new companies craved, but it also had longevity and customer service on its side. Its customer care was legendary; it had more sales and support people in the field than Cisco had employees. For the engineers and salespeople who had originated the product, joining Lucent meant going from a ten-person to a many-thousand-person sales force overnight. Within months of its deal with Lucent, Yurie was selling its product to companies like Verizon that had previously had been out of reach. Dan Plunkett, a Mercer Delta consultant who worked closely with Lucent through most of its history, believes, "Yurie was a success because the acquisition represented a specific enhancement to an existing Lucent strategy. Unlike some later acquisitions, Yurie added value to Lucent without the requirement of transforming either Lucent or Yurie into a new kind of company."

Lucent's acquisition strategy could be summed up in a single phrase as "filling the holes." Lucent's strength and market share had always been in voice transmission, and this was the part of the business that still generated the majority of its profits. The money may have been in voice, but the innovations and growth were in data transmission. Just as Fiorina had promised investors in the road show, the rebuilding had begun. Lucent's traditional customers, as well as CLECs, were all rushing to establish and upgrade their data networks, and Lucent could not begin to develop organically all the product capabilities it needed. Lucent management did not like to acknowledge how far behind it was in the data networking race or how much it would need to scramble to catch up. In Pat Russo's words, "Although we don't have an 'acquisition strategy' as such, we are open to acquiring firms that will better enable us to execute our business strategies. For instance, when we find we have a gap in a specific talent, technology, or geographic market, an acquisition may present a strong option for closing that gap. Naturally we evaluate such options with an eye toward enhancing shareholder value." There was no gap wider than that of data networking.

Why Lucent, which had a bird's-eye view on the data networking revolution, did not have the products its customers needed is a question with several answers. First, much of the data capability at AT&T had been located in NCR and been spun off with the computer company in the trivestiture. Whatever data technology existed within Bell Labs had failed to see the light of day. Lucent came out of AT&T with little capability in Internet routing, data transmission, or ATM switching, despite the fact that the last was the system the Baby Bells were adopting. "Quite frankly, we missed the last generation of data networking while a part of AT&T," admitted O'Shea. "We had all the technology to compete, but it was buried under a huge organization." Second, the Baby Bells and AT&T may have initially underestimated the size of their need for data transmission technology and how fast it would grow, sending Lucent the wrong signal. Finally, despite the fact that virtually all of the innovation in the industry was occurring in data transport, for Lucent, the money was still in voice.

The reality was that, despite the capabilities of Bell Labs, in the data market Lucent had slipped badly behind. Kim was brutally honest when he pointed out that Lucent Technologies had a lapse. "This is Bell Labs, the company that invented transistors, the premier lab in the communications field. They have 130,000 people, and they get three patents per day. In our entire company we only have 250 people, and we only have three patents for the whole thing. Lucent Technologies, with 130,000 people, had to admit that they couldn't do what we did with only 250 people. They're going to pay $1.1 billion in cash for that."

THE LARGEST TECHNOLOGY MERGER—EVER

When Lucent acquired Ascend Communications in 1999, it was not a diamond in the rough, a little-known start-up with untested technology, a few executives, and even fewer customers. Rather, Lucent paid $24 billion for the number-one vendor of ATM switches to phone companies, a mature ten-year-old company with more than four thousand employees. Ascend was the second-largest data networking company after Cisco and ahead of Bay Networks, a similar company that Nortel had bought the year before.

Ascend's potential was no secret. Its stock had risen tenfold between 1995 and 1997, and its annual sales had grown from $40 million to $2.2 billion in five years. But the financial alchemy that drove many of Lucent's investments became less effective as the company expanded. As Lucent grew, the size of the companies that it needed to acquire in order to have an impact on its own multiple grew commensurably. Like feeding a raging fire, there was no use in tossing on more of the kindling that had started it all off. Tiny start-ups with a dozen people did not make a ripple in a $150 billion company, which was Lucent's market capitalization in December 1998. Yet the larger the start-up, the less of an impact Lucent could make on its business; therefore the size of the boost Lucent's share price received was also diminished. There was no one in the market for an Ascend-type switch who did not know that it was a potential vendor. Ascend might sell more ATM switches as part of Lucent, but the impact was far less than it had been for some of the niche products Lucent had grabbed on to earlier. Still, there much hope that Ascend would have a profound impact on Lucent's valuation, since it was projected to grow at twice the rate of Lucent and was twice as profitable.

Two years earlier, Cascade had been the only other major company with an established market share for selling ATM switches to telecom carriers, and Ascend had acquired Cascade for $3.7 billion in 1997. (At the time, Lucent had also spoken to Cascade about an acquisition.) Only months before that deal Cascade had bought Sahara Networks. It was an acquisition sequence that bore more than a passing resemblance to Russian dolls nested inside one another. The Ascend merger with Cascade had not gone well at the start. Neither the cultures nor the product mix of the two companies had melded easily. In the turmoil created by the transaction, many key Cascade engineers and executives left, including one contingent that founded Sycamore Networks. With the defections and the stock market downturn in 1998, Ascend's stock price halved, and it was not until shortly before the Lucent deal that Ascend emerged from its difficulties. Lucent management looked on as the two companies struggled to create a single entity and to bridge the East Coast–West Coast gap, and then it jumped into the mess itself.

By 1999, Cisco's dominance in data networking looked unstoppable, and McGinn was searching for the Cisco-killer. The Lucent–Ascend combination looked as if it might just be the silver bullet that would finally challenge the Silicon Valley company's hegemony. McGinn argued that Ascend allowed Lucent for the first time to cover all the technology bases, but Lucent's competitors were predictably skeptical about the deal. Cisco CEO John Chambers had long ago pronounced ATM dead, a technology of the past to be eclipsed by IP, despite the fact that it continued to sell strongly to telephone service providers. McGinn did not agree, saying, "The truth is that IP [internet protocol] is not ready for prime time. We have avoided Elmer Gantry excesses and made a conscious effort not to create an image that is greater than the substance." No one could have doubted how he felt about Cisco when he called it "the best selling machine since IBM of the 1970s." Nortel CEO John Roth also argued that McGinn had bought the wrong technology and paid too much for it. "Twenty billion later, Lucent still needs to go shopping," he said at the time.

The real opportunity to take on Cisco in its own markets would have been with the acquisition of Juniper Networks. Lucent's tango with Juniper in late 1998 must be seen as one of its great missed opportunities. McGinn was dissatisfied with Bell Labs' ability to develop IP router capability and sent his emissaries to speak to Juniper CEO Scott Kriens about buying his company. Juniper Networks was founded in 1996 and provided the most credible threat to Cisco's hegemony in the IP router market. In their earliest conversations Lucent offered Kriens $250 to $275 million, but it would probably have taken twice that to secure the company. Kriens declined the offer and was unwilling to put forth a counteroffer, as he was scouting the IPO market with his bankers. Lucent rethought the proposal in 1999, and while it might have made an offer in the area of $1 billion, by then discussions with Kriens's bankers had progressed and McGinn told him that it would probably take $2 to $3 billion to keep Juniper from the public market. Still, Lucent had just paid ten times that for Ascend and the Ascend people were clamoring for Juniper's product, arguing that they could easily sell it. Lucent's management debated whether to be more aggressive in the pursuit of Juniper or concentrate resources on an alternative in-house product. Bell Labs was developing a competing IP product, the Packetstar IP Switch, which senior management from Bell Labs argued would eventually be superior. Others in Lucent violently disagreed, citing Bell Labs' unimpressive track record for expeditiously converting concepts into products. John Dickson, head of Lucent's Microelectronics division, later gave McGinn his take on Bell Labs: "How we take ideas to product is like amateur night. If we don't want Bell Labs to be the sole legacy or remnant of Lucent we had better get some hard alignment with business goals very fast." One business leader who had made his case to McGinn recalled, "I told him that buying Juniper was the difference between being a force in the router market and missing it entirely. Juniper should have been Lucent's first through tenth choices." McGinn listened to both sides and was eventually dissuaded by his own scientific team, who advised him to opt for the homegrown product already in development, which, if successful, would save the company billions. The Packetstar router was announced to the world with much fanfare in spring 1998 at SUPERCOMM, the annual industry gathering. Despite the enormous investment in development and barrage of publicity about the new router's future capabilities, the Packetstar never became a commercially viable product.

At one time Lucent might have bought Juniper for less than the IPO price, which itself turned out to be a fraction of what the company would soon be worth. Juniper's stock was priced at $34 for the IPO but opened at over $100 on its first day June 25, 1999. The price rose tenfold over eighteen months, from mid-1999 to late 2000, before returning to Earth with the rest of the sector. The day that Juniper went public Lucent bought Nexabit, a data company that employed IP technology, for $900 million. Far from staking Lucent's ground in the router market, the product languished and the staff departed en masse to begin another start-up.

McGinn bought Ascend for all the right reasons. He needed data networking talent, market position, and an established product. Despite an array of Bell Labs development efforts, Lucent's history with home-grown data products had been one of splashy announcements at large industry gatherings, delayed introductions, and later the withdrawal of a product that was late to market and now obsolete. One senior researcher at Bell Labs noted that the company's efforts in data networking did not have consistency or show a clear understanding of the goal (as the company tried to cover the technology waterfront, not committing itself to a certain market or protocol), and that this lack of focus led to a pattern of fits and starts. AT&T had a long-standing reputation of shunning anything that was not home grown. The not-invented-here syndrome was finally quashed by Lucent management, which made it clear that far from waiting for Bell Labs to invent or develop everything they needed, they would buy their way into the data market as quickly as possible. In 1998, Lucent needed a proven technology, satisfied customers, and an experienced sales force. Ascend looked like the solution to many of Lucent's problems, and vice versa. Ascend, in an instant, would plug some of Lucent's product holes.

From Lucent's earliest days, reaching a fevered pitch after the second anniversary of the spin-off in the fall of 1998, there was speculation that Ascend would be Lucent's number-one target. But the risks were high. Lucent's purchase of Ascend would be the largest technology acquisition to date, more than twice the size of the next-largest acquisition as measured by market capitalization. Large technology acquisitions have a sorry history with which Lucent's management was entirely familiar.

"This is a true case of the customer is always right," said Fiorina when announcing the acquisition of Ascend. "Our customers asked us to do this merger with Ascend, and we're seeing the results." The deal seemed to make perfect sense. Lucent needed data networking; Ascend needed an anchor, solid assurance that the company would be there five years hence. Executives from Ascend said that Baby Bells promised purchases on the condition that the smaller company partner with Lucent for the work. Service providers, such as the Baby Bells, coveted the technology at nascent companies but did not want to make a major purchase from a company that could be gone in six months. They drove many of Lucent's acquisitions, including this one. Lucent would be there in six months and six years. Telecom carriers were loath to invest in expensive "boxes" from companies whose names they knew only from a Web site and a couple of impressive sales calls and whose financing might evaporate in weeks. They had no intention of doing business with a company whose aftercare service consisted of a representative on the other end of an 800 number or, worse, a form for submitting problems to a Web site. Telecom carriers wanted the equipment developed by some of the most innovative minds, but they wanted the box with Lucent's name printed on the front.

Outside observers of the Lucent–Ascend deal chuckled from the stands. "If Ascend and Cascade had cultural issues I can't wait to see Lucent and Ascend," analyst Levy commented at the time. "Ascend only had one speed and they didn't down shift below fourth gear. That is not where Lucent operated." Even those closest to the deal acknowledged the mismatch. "Our culture had always been just to say yes," recalled Jeanette Symons, Ascend's cofounder and chief technology officer. "Lucent has been more conservative, like 'Let's plan what we will deliver to you five years from now.' " The two companies had widely diverging worldviews. Ascend's sales pitch had consisted of telling large service providers that they no longer needed the type of circuit switches that Lucent still so profitably sold. On a more superficial level, Lucent continued to call its acquisition "the Ascend unit," which Ascend managers say was a fatal mistake. Mory Ejabat, Ascend's legendary hard-driving CEO, never allowed that when he made acquisitions. The first thing he told his managers once the ink on a contract was dry was "Get their logo off the door." Within days the name of the old company had vanished.

Lucent–Ascend was East meets West. Ascend was Silicon Valley; Lucent was bedrock science. Lucent was 99.999 percent reliable; Ascend was lightning fast. Ascend valued brash aggressiveness; it was a hard-driving results- and reward-oriented culture with a flat organizational structure and a work force highly motivated by compensation. While Ascend employees worried about stock options and competition, their Lucent colleagues were concerned about pensions and long-term careers. One Ascend executive remembers an early meeting where the subject of "investment options" was raised. Then the Lucent team began to discuss how to allocate the funds in its 401k plans, while the Ascend team eyed one another warily. To them, options had only one meaning and retirement had nothing to do with it. Yet the Ascend folks were enthusiastic about the possibilities for change. One former Ascend manager explained, "We could not believe our good fortune. I remember a newspaper headline at the time, 'Ascend Wins Lucent Lottery.' But we went into this with our eyes open, believing we could help change Lucent. If Lucent had paid $24 billion for Ascend, they must have wanted Ascend's culture as well as its products, we reasoned; they must have wanted to change." Marc Schweig, senior vice president of sales of Lucent, remembers, "They were very explicit that this was far more than a merger—we should embrace each other's culture as fast as possible. Together we were going to create something that was far better than either of the parent companies."

The deal was an ill-fated marriage from the start. For acquisitions to be successful, the cultures of the two organizations need to meld; the people, products, and systems need to work in concert. Lucent bought Ascend for good strategic reasons. Ascend would bring experience, managers, engineers, customers, products, pipeline, and visibility. "Ascend could have been an injection of new DNA into Lucent," explains one executive close to the transaction. "Rich thought he would let Ascend people roam around and run things, a sort of Darwinian approach in which the best management talent would rise. But it turned out to be too destabilizing, and Lucent was far worse afterwards because it was so demoralized."

Lucent's values, in order of importance, can be simplified as quality, function, and timeliness. The products Lucent traditionally sold lasted a decade or more, so even if the development stage were delayed for six or nine months it would, over the length of the product cycle, be well worth the extra time to develop a superior product. Ascend's values were the same; only the order was reversed. Its customers wanted products "at the bleeding edge of technology," as Ascend sales leader Mike Hendren explained. The inconvenience posed by technical difficulties that were still being worked out, even after installation, was seen as an acceptable trade-off for getting out in front of the competition. At Ascend it was considered better to make a decision quickly, have it be wrong, and make another decision than to slow down the decision-making process to a glacial pace (as many Ascend managers have described Lucent's) in the hope of arriving at the right conclusion the first time. Yet Ascend cofounder Symons was the first to admit that the data networking world had much to learn about reliability from the legacy of circuit switching. She once lifted a telephone receiver during a conference presentation and explained that the dial tone was already there before it reached her ear. AT&T once estimated that only eighty-four out of a million phone calls have to be redialed, and this was not something, Symons pointed out, that those in the data world "could guarantee."

The issue of quality standards would remain a sticking point between the two companies. "One of the things they [acquired companies] bridled under is that Lucent's requirement is five nines [99.999 percent] reliability," recalled a former Lucent executive. "They were dealing with three-eighths reliability. It was perfectly okay in their market. That was a culture shock when the new Lucent, moving at an entirely different pace, would say to the acquired company, 'Your stuff is not good enough.' These guys would say, 'You're a bunch of bureaucrats. It doesn't make any difference. The customer doesn't care.' "

As much as its products, Lucent coveted Ascend's sales force. Lucent's sales force had little experience selling data products, and time was running out. In buying Ascend it would more than double the number of people in its data communications sales force. (The two sales forces were immediately combined.) Ascend was a marketing machine with an experienced sales force legendary for its unbridled aggressiveness. One Lucent salesman recounted his first joint sales call with an Ascend colleague. "The customer had called the meeting and was angry. There was a problem with an Ascend product that had already been installed," he remembers. The Lucent salesman cringed while the customer reeled off his list of grievances. Then his Ascend coworker apologized and began a new sales pitch. The Lucent salesperson, who was ready to send in the technical staff, was stunned when the pair emerged from the meeting with a new order for the next generation of Ascend products.

In Lucent's view, Ascend did not repair things; it simply moved on to the next product iteration, working out the problems then. Ascend came from a world where relationships with customers were relatively new; if you did not deliver, you did not see that customer again. One of Ascend's managers recalls Ejabat's two famous phrases: "If you are late, you lose" and "I would rather have an unhappy customer than no customer." This was a business strategy that Lucent simply could not stomach.

The difference in culture showed up at the first quarterly sales meeting of the merged firms to discuss why a sales target had been missed. The Lucent side of the team explained that they had underestimated demand and had not been able to get equipment shipped in time. "The Lucent management folks went on to discuss process and committees while the Ascend folks just wanted to know who was getting fired," recalled one Ascend sales manager.

Mike Hendren was the first point of contact with the Ascend team for many Lucent salespeople. Hendren, an ebullient sales manager from rural Kentucky, had serious reservations about his ability to fit into big-company life. But it took only one visit from Dan Stanzione, then head of Bell Labs, to convince him otherwise. Stanzione visited Hendren in Kentucky and made his case about the possibilities Lucent held. Hendren was "drawn to the aura of Bell Labs" and was impressed that its brilliant head turned out to be such a warm, approachable person who took the trouble to make his way to a remote southern town to woo a sales manager.

Hendren first addressed his new Lucent employees wearing sweatpants, a T-shirt, and flip-flops, and told them without a hint of jest, "If you miss your numbers one quarter, you get a call from me. The second time you get a call from human resources and you are gone." Hendren believed in hiring slow and firing fast. He explained to those who worked for him that they had a sacred responsibility to the employees of the company to meet their numbers. People, jobs, and success rested on their performance. Those who worked for Hendren remember him doing everything within his power to make them a success. "I believe in you more than you do in yourself," he told them. One new sales manager who came to Ascend from AT&T (and later found himself back at

Lucent) found Hendren's behavior baffling. "Mike Hendren was really trying to help me," he recalled. "He put me in contact with people who would help me, advised me where to spend my time, and told me that it was his obligation to make me successful. Coming from a large company, I did not know what to think. I kept wondering why he was doing this."

To Lucent's sales force, Hendren could have landed from another planet. They still chuckle (only because he is no longer at Lucent) over the speech he gave when he told them to find some change and telephone their mothers if they needed to talk to someone who cared. Many from Ascend remember their experiences nostalgically, describing them as the most intense and rewarding professional experiences possible. As one noted, "There has rarely been a collection of sales reps and SEs [sales executives] of the caliber put together by Ascend. These folks were legendary . . . and let's not leave out the fearless leader, Mike Hendren, the most out-there, over-the-top sales leader that ever pushed his guys out the airplane door without parachutes—'cuz you gonna *learn* how to fly, damn it!" But not everyone was impressed. "People were offended by a guy who showed up in sandals, was disconnected from company protocol, and said exactly what he thought," Lance Boxer remembers. But McGinn was a big supporter. He viewed Hendren as an agent of change and encouraged his efforts to create a more aggressive sales force.

At Ascend, most members of the sales force had engineering backgrounds and needed little support from technical people to help make their sales. At Lucent things were different. Lucent sold large, complex communications systems that required a fleet of engineers to integrate and install. Layers of technical people who worked with customers to design systems and make the sale backed up the salespeople. Hendren was appalled by the number of Lucent people it took to cover a customer. He did not win any new friends among the Lucent sales force by repeatedly sending out this message in colorful and graphic language. Ascend people did not think Lucent got it, and the reverse also was certainly true.

Ascend's management made it clear that it wanted to make money. They loved the workaholic culture in which they had been immersed for most of their careers and hoped to bring it to Lucent. This would prove far easier said than done. Lucent's own culture was undergoing such a headlong transformation that it was a challenge to incorporate new elements into the mix. While McGinn publicly insisted that the two companies were a perfect fit, the litany of adjectives senior Lucent managers used in private to describe their Ascend counterparts ranged from the more flattering "aggressive" to the more colorful "brutish, rude, unsophisticated cowboys" and "Valley greed personified." Lucent types, up through the highest levels of management, found Ascend's obsession with compensation and reward appalling. They saw it as a mercenary approach to employment, with no value placed upon institutional loyalty or job security. One senior Lucent officer remembers a breakfast in a Marriott hotel on a visit to California to see Ascend officials. "A group of them [Ascend staff] sat at the next table and for the entire meal talked about nothing but their stock options." Resentments abounded. Lucent staff believed that Ascend employees had all become wealthy in the buyout, and now their new coworkers held larger chunks of Lucent stock than most lifetime Lucent employees. One Lucent senior executive suggested that his staff's reaction to this depended on how threatened they felt by the Ascend dynamos unleashed in their midst.

Imagine how some of these transactions must have looked to the tens of thousands of tenured longtime Lucent/AT&T employees. In some cases, companies with untested management, no earnings, and without a product could convert their ethereal promise into billions of dollars of real Lucent stock. These arrivistes became millionaires, and in a few cases billionaires, and overnight were airlifted to the top of the corporate heap without the life-consuming tedium of actually having to scale the corporate ladder. They sat in offices beside Lucent managers, often with similar educational backgrounds and ages but fundamentally different aspirations and a net worth ten, one hundred, or even one thousand times that of the lifer in the next office. A number of people who had been with Bell Labs or AT&T before joining the start-ups that were acquired by Lucent savored their triumphant return complete with bursting bank accounts and vice president titles. Despite having the same job descriptions, they came from different worlds. McGinn loved this. He often mentioned that the company had a few billionaires and many deca-millionaires. He viewed this phenomenal wealth creation as a badge of success.

McGinn liked to brag that Lucent completed thirty-eight acquisitions in five years, but half of its investment was made in Ascend. Lucent paid a fortune for Ascend, and its challenge, as always, was to hold on to the talent it had bought. McGinn had no illusions about the essential ingredient in an acquisition: "If you don't win the hearts and minds of the employees then all you've bought is a shell." Lucent was having a difficult enough time holding on to its homegrown talent, now inundated with job offers, let alone skeptical outsiders. "We all had opportunities outside of Lucent," says Sam Mathan, former vice president at Ascend, now CEO of telecom start-up Amber Networks. "For me personally, I like to build things. I belong at a start-up, not a big company." Mathan's views reflect those of many of his Ascend colleagues who, once paid off by Lucent, could not wait to take up the challenge of another start-up. At the time of the merger, analyst Levy observed, "Lucent's management was not very lucid in explaining how it intended to prevent the 'brain drain' from continuing post a Lucent takeover and insisted that the two companies' cultures were similar. With all due respect to the accomplishments of the current Lucent management team in the last three years—it's not easy taking a Bell System mentality and becoming entrepreneurial—we find it hard to believe that Lucent has anywhere near the corporate culture that Ascend has."

But McGinn recognized talent and drive and had none of the old AT&T wariness about yielding power to outsiders. He hoped that Ascend's leaders would be able to bring some of their culture to Lucent. But none of them would stay. Neither Ejabat nor Symons had promised to stay, informing McGinn from the start that they would remain only for a transition period. Ejabat had emigrated to the United States from Iran in 1976 and had teamed up with Symons at Ascend in 1990. They were the yin and yang of a technology company, Symons said, noting, "Mory has an excellent sense of balance between the various disciplines in the company, focusing on customers and execution. Meanwhile, I really focus on delivery of the products to the customers to meet the demands. It is a good balance between us." The very day the Ascend deal officially closed, the company's erstwhile leaders were incorporating a new business, Zhone Technologies. Within six months Ejabat had raised half a billion dollars and had thrown himself into a new venture. "I love to work. I might be crazy, but I love it," he says. "I work all my life. I cannot stop." Personal loyalties at Ascend ran deep, and this did not send an inspiring message to the troops. For Ejabat it was only natural that he would shy away from an opportunity at Lucent, where he would be forced to slot into the massive organization. As he often told his staff, "I like to work [directly] for the Board."

Ascend people did not find working at Lucent fun. Lucent compensated its employees less generously, and it was located in New Jersey. These were not selling points for Ascend. What Ascend employees saw when they arrived was a slower-moving company steeped in the corporate ways of long meetings, consensus-building excursions, and a Byzantine corporate matrix. In the early days of the merger, Ascend salespeople who were out in the field would telephone Pat Russo with their problems, going over the heads of many layers of middle management. Russo was receptive, interested in hearing the message from the field, unfiltered. It had never occurred to anyone from Ascend that calling your boss's boss could be a problem, because Ascend's organizational structure had been so flat that the problem rarely arose. Yet the Ascend salespeople were warned by their Lucent colleagues that such a thing was never done. Before long those managers who had been bypassed to reach Russo made it clear that Lucent was a company with established channels of communication. Soon the former Ascenders learned to conform or found their way to the door.

Start-up technology organizations, even the largest ones, are often idiosyncratic companies dominated by the personality and ethos of their founders and leaders. With their short histories, personal loyalties are focused on individuals rather than institutions. Employees in these organizations thrive on the intensity of the "do-or-die" atmosphere. They love the climate of uncertainty and risk taking in which they work. For Ascend employees, arriving at Lucent was like landing on alien soil; most had never worked for a large corporation and were entirely unprepared for the structure and bureaucracy such organizations entail. They saw influential managers, whose products may have sold well in the past, protecting their fiefdoms in the fear that new products might cannibalize their existing product lines, heresy in an industry that ate its young. All of the good reasons for working for a big company—stability, good pay, impressive titles, and institutional loyalty—were simply not relevant.

Lucent management believed that much of the talent that came through its acquisitions would remain with the company, finding new challenges in a large organization. It was a fanciful notion that did not come to pass. Most of the entrepreneurs who came to Lucent were people who would not have chosen Lucent as their employer or perhaps even included it among their first ten choices. Even many of those who were determined to stay describe being worn down by Lucent's glacial decision-making process and bureaucracy. Almost every executive from Ascend was gone within a year, as soon as their shares vested. The damage from this turnover permeated Lucent's original sales organization, as McGinn had put many Ascend leaders in charge. Any expectation that Lucent could swallow Ascend whole and emerge unscathed proved unrealistic. "Lucent had an antibody reaction to Ascend," explained a senior Ascend executive, who spent his requisite year with the new company, then left.

"The problem with most people from Lucent was that they had never known fear," recalled a sales executive who came from a start-up and returned to one after his experience at Lucent. They had never struggled in an environment where management was afraid of not making a payroll, had never feared having a competitor beat them to the finish line on a product and then as a result going out of business. Veterans of high-tech start-ups all over America knew fear. They knew that nasty feeling in the pit of their stomachs that with a single misstep any quarter could be their last. Their staff labored until midnight because they feared failure and loved what they were doing. They skipped touchy-feely group-building sessions because groups get built much more solidly by immersion in a common goal than by days spent locked in a sterile hotel room, and because in this world there simply is not the luxury of time. As much as McGinn might have tried to create entrepreneurship, with its 10 percent share of the global communications equipment market Lucent did not behave like a company desperate for the next cash infusion from a venture capitalist.

One Lucent vice president who joined through the acquisition of Ascend and was initially committed to a career with Lucent recalls the moment of truth in his short Lucent tenure. He had been called to a meeting, of which he was already beginning to believe that there were far too many, and sat through a lengthy discussion the purpose of which seemed to be to avoid conclusion or action. When it became clear that nothing would come from the meeting, the longtime Lucent manager who had called it stood up, smiled, and without a trace of frustration said, "Just want to make sure I can report back that nothing was decided."

The Ascend deal was a short-term success and a longer-term failure. At the time a Lucent spokesman accurately pointed out, "Four years ago, we weren't even in the data-networking business—now that area is seeing 40 percent growth every quarter for the last four quarters." Lucent did get a solid foothold in the data networking business overnight with the Ascend purchase. However, between 1999 and 2001, Lucent's share of the ATM market steadily eroded, and by 2002 it was in second place, trailing Nortel. One of the major goals of the Ascend purchase had been to acquire an experienced sales force with a dynamic culture and a proven track record. But it is hard to find anyone from either the Lucent side or the Ascend team who will argue that the two companies were anything but an abysmal fit.

CAVEAT EMPTOR

As its own stock price was on a tear, Lucent could afford to engage in bidding wars with its competitors, like wealthy matrons in a Sotheby's showroom driving up the prices against one another. The price of each deal would be set by the inflated price of the most recent deal, rather than through a realistic appraisal of the future cash flows emanating from the acquisition.

Nothing was bought and dismantled faster than the Israeli start-up Chromatis. A mere ten months passed between the time Lucent took possession of the optical networking company in August 2001 and shut it down. Operating in one of the fastest-growing segments of the optical networking market, Chromatis had developed technologies for switching Internet traffic in densely populated metropolitan areas. Its product would ease congestion on crowded city networks. Orni Petruschika and Rafi Gidron founded the company in mid-1997, and Lucent Venture Partners made an investment in 1999, although after brief conversations, it backed away from the idea of buying the company outright. Lucent management agreed at the time that Chromatis's technology looked promising, but it was too early for a larger investment.

A year later, some of the Chromatis team was interested in cashing out, while others felt that they should wait and develop the product further and that in time a public offering would be a realistic possibility. Chromatis did not approach Lucent as a potential buyer because the relationship between the two companies had soured when Lucent had lost interest the year before and negotiations for an original equipment manufacturer (OEM) agreement between the companies had ended badly. Chromatis's management felt that Lucent had behaved unprofessionally, and ill feeling lingered.

By May 2000 Chromatis still had no customers and no revenues, but it had an interested suitor. When Chromatis began serious discussions with the prospective buyer, its investment banker telephoned the handful of other companies that would want its product and could afford the price. One of the unspoken goals of many telecom investments was keeping new and untested technology out of the hands of the competition. Few companies wanted to risk seeing their competitors buying up a small research effort only to find that it was the "killer app" that would dominate the market next year.

Informed that Chromatis had an interested buyer, Lucent discovered that it, too, was interested. Lucent had come to believe that it could capture the metro market the way Nortel had recently dominated the optical switching space and that Chromatis's product would be key to this assault. Journalists following the telecom sector speculated at the time that this deal would be a real threat to Cisco, which had recently bought a company with similar capabilities that had not yet panned out, and would trigger Nortel to rush out and buy its own metro product. By the time Lucent telephoned Chromatis, the smaller firm's management was already in possession of the term sheets for a deal it hoped to close elsewhere. Lucent began the conversation by acknowledging the bad feeling and saying that despite the fact that it had turned down the buying opportunity nine months earlier, it was genuinely interested now.

Robert Barron, the company's president, told Lucent a deal was impossible: the timetable was too short, and it had arrived too late. Lucent insisted on being given a chance, so Barron set out three conditions under which Chromatis would agree. First, as the others had already proposed their terms, the deal had to be done in four days. Chromatis had no intention of losing the bird in the hand for some vague expression of Lucent's interest. Second, the financial offer had to exceed the one on the table. And, finally, once the transaction was completed, Chromatis had to be left alone to finish its product. Lucent might buy Chromatis, but it had to agree not to interfere.

Lucent accepted these conditions, saying that it thought it was better that Chromatis finish its development process unimpeded, as experience had shown that when it interfered, things did not work out as well. Lucent management arrived by helicopter at Chromatis's Hendon, Virginia, offices that same evening to begin discussions. In four days the deal was done.

In January 2000 a venture capital firm had made a $10 million investment in 10 percent of Chromatis. Now, with a bidding war and Cisco's recent $6.9 billion purchase of Cerent (a company in a similar marketplace) as the yardstick, that same 10 percent stake was valued at $450 million. The value of Chromatis had risen forty-five-fold in a matter of months in a market that had lost its way. Lucent paid $4.5 billion for a company that had yet to make a sale.

Chromatis's management sold its company to Lucent for much the same reasons that had prompted the thirty-seven preceding Lucent acquisitions. The price was spectacular, but, more important, Chromatis had believed the caliber of Lucent's manufacturing and sales was such that its dreams for its product would be realized. And, like all those before it, Chromatis was swayed by the words of its own investment bankers, who said, "Lucent's stock is as good as currency."

The Chromatis story ended badly. Within thirty days of completing the deal, everyone who had negotiated from the Lucent side had left the company, and the terms of the deal were soon ignored. Barron, Gidron, and Petruschika, who were to be left alone to run the new investment, were soon drafted by McGinn into one of Lucent's main divisions. Chromatis was shifted back and forth between different departments within Lucent, and its specialized sales team was eventually merged into a larger sales group. By 2001 Lucent was focused on its major customers and bread-and-butter operations, and that summer it shut Chromatis down. Speculative next-generation technology, even with its massive potential market, was no longer in the cards. Lucent wrote off all of its investment in Chromatis. With no product, no customers, no technology or synergies, Chromatis simply dissolved.

By way of acquisition, some of the world's best talent walked through Lucent's doors and back out again. In retrospect, this must be seen as a missed opportunity of monumental proportions. Many who emerged from this experience describe Lucent as far too slow-moving to provide a challenging opportunity to those with entrepreneurial aspirations. For them, the rhetoric did not match the reality. They speak of their disappointment at not being able to realize the ambitions they brought to Lucent and watching the companies they had built disintegrate.

ENRON, HOME DEPOT, AND LUCENT

Confident of Lucent's prospects, McGinn staked his ground and put in place hugely ambitious growth targets. Lucent would not be a "safe stock," one that widows and orphans had tucked into their portfolios by well-meaning bank managers. Plodding along, shedding dividends with no prospects of a heart-fluttering move in the share price, was not what McGinn had in mind. "Why would someone want to create a fairly good outcome or a reasonably good company?" he asked. "In my view, I and we, the leadership of the company, had the opportunity and the responsibility to do absolutely the best job we were capable of doing. We had the responsibility to eliminate the cynicism and notions of best-efforts performance, which were present in the minds of some. We had the responsibility to take prudent risks, not reckless risks, to reinvent the company as it was being formed. As leaders, we had to both teach new skills and reset expectations while we reinforced strong positive values. I for one would never want to be involved with any effort where we aspire to reasonably good performance. The job of leadership is to inspire and to lead. Many wonderful things happened because we had high expectations of ourselves. The clock did run out, in the sense of the entire market imploding, but a lot of people improved their skills greatly because that was expected of them."

Between 1997 and 1998, Lucent's total revenues increased by 14 percent and profits were up by 49 percent. Industry growth was expected to continue at 14 percent a year, and McGinn promised his investors that going forward, Lucent would best the industry pace by between 3 percent and 5 percent. This was what McGinn told the outside world, but privately his team aspired to an even faster pace and hoped to grow Lucent's $30 billion business by almost 20 percent between 1998 and 1999, a target not out of line with what many of his competitors were pledging. But while the torrential spending in the telecommunications industry might have made this possible, history was not on his side. At an annual growth rate of 15 percent, Lucent would double in size every five years. Lucent management grew fond of saying that there were only three companies in the world that were growing at 20 percent a year and had at least $30 billion in sales: Enron, Home Depot, and Lucent. Plenty of small companies are able to expand their businesses at such a pace, but few industry leaders can grow this fast for extended periods. Russo was wary about the push for growth, she remembers. She says, "Rich was pushing the business very hard for growth and believed, given what was going on in the market and the opportunity that we had, that we should have been meeting the growth expectations." Although in the short run McGinn's projections looked achievable, he was venturing out on a limb. Either he would be conducting only the second job search of his career or publishers would line up to offer hefty advances for his autobiography. With such an ambitious goal, it was hard to see how he might locate the middle ground.

Setting audacious growth targets had become seemingly irresistible across many industries. CEOs all over the country were popping up with such claims. Good times can presage more good times, and as the booming economy set new records for both duration and magnitude, more CEOs thought their companies could do the same. With the added incentive created by stock-based compensation, any CEO achieving such a goal, even for a brief time, stood a good chance of emigrating from the land of the wealthy to that of the superrich. Sizable companies that grow by 15 percent per year, as Carol Loomis of *Fortune* points out, make the all-star team, and their CEOs are appropriately feted. This in turn encourages others to put similar targets into place that, as she notes, create the incentive for uneconomical behavior focused exclusively on short-term profitability. Few managers who promise such growth rates ever manage to achieve them, but many inflict incalculable damage on their companies in their attempt to do so. At Lucent this push for growth would have a more sinister outcome as a group of salespeople ventured into illegal and unethical territory, bringing down the wrath of the SEC on themselves and the corporate body.

Steven Levy conceded that fantasy had become a factor in investing as he attempted to explain Lucent's valuation in early 1998: "In setting our new $90 price target for Lucent we have essentially thrown in the towel on any traditional valuation parameters. . . . Although we do this reluctantly, given our long history of following technology stocks . . . we also recognize that today investors are rewarding highly liquid, relatively rapidly growing companies with somewhat astronomical valuations." From Federal Reserve chairman Alan Greenspan on down, there were authoritative voices arguing that levels of productivity, particularly in the technology industries—the likes of which had never been witnessed before—were now solidly in place.

Despite decades of evidence, the stock market no longer wanted to believe that telecommunications was a cyclical, capital-intensive business that grew somewhat faster, but not two or three times faster, than the economy as a whole. Companies that could legitimately use the words "Internet," "optics," or "network" in their annual reports soon found their P/E multiple had doubled or tripled. Few chose to view this unprecedented growth as simply a phase in a cycle that would soon come to an end. McGinn was completely in step with the times when he ask his troops to believe in miracles, in growth that had never been seen before. As he put it, "I wanted everyone to suspend their disbelief. I wanted to set goals that we didn't know how to achieve."

Caution, and its soul mate fear, had once haunted an investing public convinced that it had almost as much to lose as to gain. But over the course of the 1990s a confluence of events, including sustained low interest rates, abundant capital seeking excess returns, and a belief that in the long run stocks were guaranteed to provide a superior return, blunted investors' caution. It was the Internet that swept away any lingering concerns about inflated stock values by convincing the investing public that this time was different. The Internet was believed to be something more than simply another step in the development we used to call "progress." Rather, it was a transformation so great that economic outcomes, be they in the form of personal wealth, lower national unemployment levels, or worker productivity once thought to be impossible or at least highly unlikely, suddenly seemed probable. Investors believed that when markets declined they would not do so for extended periods of time, and that those with patience and staying power would eventually be rewarded by even higher stock prices. The diminution of fear was a boon to the price of a stock such as Lucent's. Businesses in which the CEOs drove their companies faster and faster in ever-more-uncertain conditions might have been given lower valuations in a more cautious era. But the normal logic had been reversed, and companies like Lucent were seen not as risks but rather as great opportunities for growth.

The pressure Lucent found itself under had begun more than a decade earlier, long before its birth was even contemplated. During the 1980s, American investors began to worship at the altar of shareholder value. This meant that it was the responsibility of the CEO and his executives to manage their company in such a way that over time the owners of the company (i.e., its shareholders) earned the maximum return, or the maximum present value of its future cash flow. "Over time" did not mean one quarter, one month, or one afternoon, but rather year after year. But the notion of maximizing shareholder value altered radically over the decade, and by the late 1990s it had been grossly distorted to mean increasing the stock price of a company regardless of its cash flow. Markets soon forgot that stock prices are the result and not the cause of value creation.

To ensure that they kept their focus, managers' compensation was tied to their company's stock price. Thus for CEOs all over America stock prices came to serve as a daily report card, an assessment of the state of their business by their millions of owners. Ideally a company's stock price should be a *reflection* of its growth and earnings, but in a convoluted turn of events companies began to manipulate their performance and manage the information they reported in an effort to set their own stock price.

Lucent was far from alone in feeling the pressure to produce revenue growth that outstripped that of both the industry and the economy. The perceived expectations of investors were enough to induce a CEO to "make the numbers" and display a smooth upward trend in growth rather than risk having to "preannounce" bad news and bring on a precipitous drop in share price. Setting and meeting market expectations subsumed all other goals. The effort that goes into legally manipulating the quarterly accounting figures of a major corporation is an enormous nonproductive expenditure of executive time. And the numbers, once revealed, have no predictive value, as they are little more than an accounting artifice. This is an opportunity cost, but Harris Collingwood, a senior editor at the *Harvard Business Review,* suggests that the quarterly manipulation of information that was widely expected had very real costs as well: "The fetishistic attention to an almost meaningless indicator might be cause for nothing more than amusement, except for one thing: the earnings game does actual harm. It distorts corporate decision-making. It reduces securities analysis and investing to a guessing contest. It compromises the integrity of corporate audits. Ultimately, it undermines the capital markets."

Once Lucent became a public company, it began the "dance of the analyst." After it became clear that Lucent was a growth stock, the dance became fast and furious. This is an almost farcical process by which management gives analysts hints about what to expect (often in private conversations); analysts in turn tell management what they expect, and then grade them on meeting those expectations, which of course management established in the first place. There is a veneer of objectivity in the form of the computer models that analysts tout, but there can be no doubt that their expectations for a company's performance come directly from the company's management. In a language that has grown up between analysts and those who speak to them, full of hints and nuances that keep both parties from running afoul of the SEC, companies painstakingly delineate the numbers they expect to produce. Analysts take these carefully coded messages, the content of which is proscribed by the SEC rules that govern the trafficking of insider information, and decode them for a public that has not the time or the inclination—or the possibility—to sit in on conference calls in which executives drop their carefully worded hints.

Each quarter, McGinn would spout the enthusiastic predictions that investors loved to hear, analysts would run back to their desks to write rosy reports parroting those predictions, and investors would buy Lucent's shares. Lucent would then report numbers that were at or above "expectations," and everyone would look good. Management had been right, Wall Street had been right, and the investors who continued to profit saw few reasons to challenge a system that appeared to be working. Later McGinn would warn that expectations would not be met, analysts would revise, investors would sell, and the dance would begin again, quarter after quarter, with only a rare independent voice heard above the crowd. Arthur Levitt, while still chairman of the SEC, decried this system long before it became fashionable to do so: "I can only point to what I see as a web of dysfunctional relationships—where analysts develop models to gauge a company's earnings but rely heavily on a company's guidance; where companies' reported results are tailored more for the benefit of consensus estimates than to the reality of the ups and downs of business; where companies work to lower expectations when they fully expect they'll beat the estimates; and where the analyst attempts to walk the tightrope of fairly assessing a company's performance without upsetting his firm's investment banking relationships."

Not every analyst was listening to Lucent management. Tony Langham, a telecom analyst with Smith Barney, was not impressed with Lucent's management from the start. At his first meeting with Schacht, the new CEO had declared that Lucent was number two in the world in wireless infrastructure. But Langham had the figures to prove that Lucent was number three, trailing Ericsson and Motorola, and from that point on he listened to Lucent with a skeptical ear. Langham likes to wallow around in the dirty details of accounting. In February 1997, after studying Lucent's quarterly filings with the SEC, he was troubled by what he found. No one had published an ill word about Lucent until he detonated his bomb. "Lucent is using several accounting practices which are less conservative than other telecommunications companies we follow," Langham began. He had trouble with Lucent's pension credits, reserve reversals, reserves for vendor financing, and decline in gross margins. But perhaps his most damning criticism came from what he saw as Lucent's "carefully crafted approach" to Wall Street. Most firms, he felt, had been co-opted by the flood of investment banking business that flowed out of Lucent. With such a lucrative client, most analysts were caught in a conflict of interest from which there was no escape. In 1997, Smith Barney did no banking business with Lucent and Langham was free to speak uninhibited. "We believe it is important for investors to understand the way information is being disseminated by the company. We have been remarkably frustrated in our attempts to do our job of analysis," Langham told his investors. Lucent was not forthcoming. "Lucent put on this incredible show, it was very professional," Langham recalls. "Once a year they had all the analysts in. There was a huge tent, they had handouts on everything, all their top people were there, and they literally had a red carpet. But they wouldn't answer the tough questions."

Lucent did not take this attack lying down. The day after Langham disclosed his report; Schacht issued a pointed rebuttal in *The Wall Street Journal:* "While every analyst has the right to assess our business and issue recommendations as they see fit, there is positively no reason to question the integrity of our financial practices."

To keep the analyst community satisfied and please his investors while outperforming his competitors, McGinn would not only have to scale the 15 percent handily, he would have to do it with smooth earnings that could be communicated to the market with no surprises. Later, in conversations with the Hewlett-Packard board, Fiorina criticized what she saw as McGinn's unrealistic promises to Wall Street, assuring the board that she would not make the same mistake. But in 1998 she helped him achieve these goals.

Executives, analysts, and investors, in fact anyone with common sense, knows that businesses do not run smoothly quarter in and quarter out, particularly in industries with heavy R&D expenditure (where investments made now are recouped with income earned many years later) and products that carry multimillion-dollar price tags. Lucent's business was lumpy; it consisted of contracts for entire communications systems that might take years to build and install, and in many cases would have price tags in excess of $1 billion. The notion that in every three-month period earnings per share would rise by a steady twenty to thirty cents, producing an unending, unbroken stream of record quarterly earnings, was as fanciful as it was misleading. Yet all sides willingly engaged in a mutual fantasy that reported sales and earnings would rise smoothly and nothing untoward would need to be done to accomplish this feat.

Predictability became a valued trait, something for which shareholders (through inflated price/earnings multiples) would readily pay more. This led to a tacit agreement by companies all over the United States to "manage" earnings, to reformulate them within the letter of the law in such a way that no jagged edges would show. CEOs were often frustrated by being punished for long-term investments that had a negative impact on short-term earnings. When Nortel acquired Bay Networks in the summer of 1998, CEO John Roth believed he was taking a bold step in moving his voice-centered company into the data market. Investors disagreed, punishing Nortel stock to the tune of $3 billion in a single day for not delivering on Wall Street's expectations. Roth asked an institutional investor what he wanted from Nortel. The money manager told Nortel's CEO that he wanted a nice, predictable company. "With that kind of thinking, you'll be able to predict when I'll be bankrupt," Roth said. "Right," said the investor, "but then I'll know when to sell." One can only speculate on the fortunes that would have befallen a company that chose not to play the game as prescribed by the new rules.

Technology companies unwilling to promise and deliver this growth in revenues would be unable to make the acquisitions that would underpin future growth. A higher stock price could be achieved at a more leisurely pace. Promise investors moderate growth, exceed their expectations, and over time the stock price will rise. But in an industry where dominant market share in a new technology can be gained or lost in a mere six months, the inability to buy the needed technology can mean missing an entire product cycle.

Valued employees might also be gained or lost with the rise or fall of a company's stock price. Lucent's workforce was not historically the type that would abandon comfortable pension plans to join thinly capitalized start-ups. However, times had changed, and as the chance of participating in an IPO or being showered with stock options increased, the lure became irresistible. At first a few Lucent employees left. When they reported back to their colleagues on the advantages of life outside a large corporation, the trickle became a flood. Without its own rapidly rising stock price and a generous options policy, it would have been impossible for Lucent to hold on to incumbent talent. One new manager from an acquired start-up remembers that a large portion of Pat Russo's first presentation to the now-united management team focused not on Lucent's strategy or the new businesses' place in the larger organization, but rather on the reasons why they could expect Lucent's share price to continue to rise. Lucent's share price had become headline news. Morale soared as the stock price doubled again in 1998.

The self-image that McGinn projected for the company was exciting and highly motivating to employees who had gone to work for a regulated utility. Schacht had represented more measured change, with his "aw, shucks" sense of humor and his frequent references to his experiences in the truck engine industry or lessons learned from what he often referred to as "capital goods." McGinn invariably focused on the fast-moving industries growing up in Silicon Valley. Talent would be attracted by talent. Unless Lucent was perceived to be a scorching-hot stock, industry talent would flee westward or to the start-ups that now littered New Jersey. Later, when the stock did falter, competitors wasted no time and swooped in like vultures.

Cisco CEO John Chambers and Rich McGinn were never best friends, but until 1998 both had remained open-minded about the possibility of their companies forging some sort of working partnership. McGinn made the first overture and flew out to California with Bell Labs executive vice president Bill O'Shea to meet Chambers at Cisco headquarters. McGinn had hoped that the two companies might fill each other's product and customer gaps. Lucent was strong in voice transmission, Cisco in data. Lucent had the largest market share in selling to service providers, Cisco in selling to enterprises. Surely, McGinn reasoned, the two companies had something to offer each other. It was even possible that a merger might come out of the discussions. McGinn and

O'Shea arrived on "Take Your Child to Work Day" and found Chambers in the lobby of his headquarters dispensing ice cream to the children of his employees. The two teams got along well, and they spent the day cloistered in Chambers's conference room discussing where their markets were headed and the general strategy of their businesses. After a day of meetings they agreed to put together a joint engineering team that would offer an integrated communications systems to a few customers. It was a toe in the water, a first step in seeing how the two companies might work together. They asked AT&T and Sprint if they would be interested in seeing a joint proposal for systems that integrated the products of the two companies. Both phone companies were highly enthusiastic. The engineering teams began to coordinate, but the CEOs fell out of communication. In a matter of a few months, relations between the CEOs had grown frosty. "We've talked about it very aggressively—it's just not going to happen," Chambers said after the meetings with McGinn. "Our products are overlapping too much and our strategies are overlapping too much." McGinn saw it differently. From Lucent's point of view, Chambers was not as interested in a partnership as he had initially indicated and McGinn was bitter about the outcome of their discussions. Chambers publicly questioned Lucent's business strategy and product decisions. McGinn set up Chambers within the company as the man to beat, and his dislike certainly became personal. Later, McGinn's own managers and employees would agree with Schacht when he said, "I think there's been an increasing preoccupation with Cisco, and it hasn't helped much."

No one had joined AT&T believing fabulous riches would follow, but by 1998 Lucent employees could be forgiven for wondering if such a possibility was now within their grasp. Chambers may have referred to Lucent as an old-world company, but in Murray Hill few doubted that they were on the cutting edge. By 1998, Lucent's stock price had risen 400 percent since the IPO; maybe the kind of wealth reported in business magazines *could* be created outside of California. But Lucent would have to make its numbers.

Companies like Lucent with the highest P/E multiples, and thus the greatest expectations built into their prices, risk the most dramatic downdrafts if they disappoint investors. By 1998, Lucent's P/E had risen above 35, or almost twice the industry average. Avoiding disappointment became a mind-numbing and tortuous process that soon absorbed the attentions of the entire organization, diverting energy away from more productive activity. If a CEO needs to give his investors good news every quarter, then his CFO and the entire sales structure must focus on meeting quarterly targets, be it by taking excess funds and putting them into reserve for a rainy quarter or by taking uneconomic action to pull revenues into an earlier quarter. At Lucent the distorting pressure emanating from the stock market reached deep into the organization, deforming values and affecting the behavior of employees at all levels. As the stakes rose with the stock price, the process of coaxing information out of the organization became even more difficult. McGinn explained it when he asked his managers what they were going to do for each of their pieces of the business: "Tell me what you're going to do, tell me what we can absolutely count on, tell me in a fulsome way you know what sales, what product is going to be delivered, availability of product—don't tell me it's a great opportunity but we don't have any components because the market is too tight around the world."

Some of the most spectacular success stories of the decade set the goalposts, and any CEO who was not reaching for them was not considered to be doing his or her job. For a few, this would be an invitation to deception and fraud, but for those in the mainstream it was an invitation to test the elasticity of generally accepted accounting principles, to stretch their legal limits. When in the normal course of business a less-than-stellar quarter occurred and CEOs found themselves unable to keep their promises, the temptation to bury the truth under increasingly convoluted and aggressive accounting assumptions became overwhelming. This is not to say that Lucent's stock would not have risen anyway over this period. The exceptional operating performance of the business would have warranted a significant rise in the stock price. McGinn's assurances to investors just provided a little more octane.

Every quarter, McGinn underpromised and overdelivered, and when he reported to shareholders the final results for the fiscal year 1998, revenues had increased an impressive 14 percent and earnings (something seen less often in the "new economy") had exploded by 52 percent. The good news had come from everywhere—the Baby Bells, CLECs, and especially international sales. Lucent was stealing market share from companies such as Motorola (one of its old icons), Nortel, and Ericsson. Its cost structure was becoming more attractive and its product mix more profitable.

But Levy was becoming nervous. Lucent beat analysts' estimates every quarter, and when McGinn met with the messengers of Wall Street at the analysts conference at Murray Hill on September 17, 1998, he wowed his listeners with his most ambitions plans. McGinn's message was this: "We have examined our addressable market, which is a $300 billion market, and have determined that it is likely to grow at 14.6 percent. Our growth will outpace the market, and we expect to see 20 percent sustained revenue growth with earnings per share growth of 35 percent." McGinn's audience was ecstatic; these were the words investors longed to hear. But to Levy, "that's when the arrogance began." Fourteen point six percent, not 14 percent, not 14.5 percent. Lucent believed that it could predict the growth rate of telecommunications, an over–$300 billion marketplace, to within a tenth of a percentage point. Levy returned to his office and scrolled through files and files of Lucent spreadsheets on his computer. Despite the grandiose promises, Lucent had had only one quarter when growth exceeded 20 percent, in March 1998. But these numbers included the purchase of Octel, a September 1997 purchase of a messaging company. Without Octel, Lucent had never had a 20 percent quarter, and now it was promising a 20 percent year. Levy did not like the sound of it. "They were saying that although they were a much larger company, they were going to grow even faster and, added to that, outpace the growth rate of their market by an even wider margin," he observed. And where was the credibility for this argument? In each of the previous two years Lucent had exceeded its stated growth targets. But in Levy's mind these targets had been reasonable and increased gradually over the company's first three years. The leap to 20 percent would mean playing in an entirely different league. Levy believed Lucent had stepped over the line, making promises it could not keep. So he cautioned his investors to sit on the sidelines. He was uncomfortable but would need further evidence before he could sound the alarm. Levy knew that McGinn had set himself and his management a heroic task that would succeed only under perfect conditions.

CHAPTER SEVEN   1999

1999 Stock price   High: $64.48   Low: $36.00

> *Being driven by growth, growth, growth, and*
> *maybe a little profitability after it, doesn't yield*
> *the best results.*
>
> Deborah Hopkins
> former Lucent CFO

Lucent would lead the networking revolution, McGinn had declared during the company's earliest days. In two years, he had successfully pieced together the business he had promised to create, and with its purchase of Ascend, McGinn believed that Lucent now had the product mix and marketing talent to dominate almost every market sector in which it competed. McGinn had pushed his troops, reset expectations, and presented a vision of profitability and growth that no one else in the organization would have dared. It was a brave and, his critics would later say, foolhardy dream, but without it Lucent might have been the unremarkable offspring of a failing telecom giant, noteworthy only for the breadth of its shareholder base. McGinn had dared to dream, and he had dared to fail.

In January 1999, Lucent was forced to reveal that all was not as it should be. Wall Street was expecting Lucent's revenues to be a robust $10 billion for the first quarter of fiscal year 1999 (beginning October

1998); instead the company reported sales of $9.2 billion. Although it was an all-time record for Lucent, the figure reflected only a 6 percent increase over the first quarter of 1998.

McGinn made excuses. His sales team had deals in the works that would have put them over the top, he said. But on the last weekend of the month, they had decided that in the name of "appropriately conservative accounting" practices, they would wait until the following quarter to recognize the revenue. It represented only a handful of contracts, much of it in software sales. Without this technicality, Lucent would have had a "blow-away quarter," McGinn stressed. There was no weakness in the business, no letup in the raging demand for Lucent's products, he told his investors. In short, there was nothing to worry about. Although profits were up strongly, the market was unimpressed. Lucent's shares were priced like a growth stock and unless they behaved like one, the share price would be firmly trounced. McGinn had promised that Lucent would outpace the industry, but now it showed signs of struggling just to keep up. The first quarter's numbers were always Lucent's best, so the disappointment was even more acute. McGinn was undaunted; the missing revenue would show up in the following two quarters. January, which was the beginning of the company's second fiscal quarter, already looked strong. Lucent, he insisted, would increase sales by more than 30 percent for the second quarter and would reach its target of 18 percent to 20 percent revenue growth for the year.

McGinn had chosen Lucent's path and was not one to step back from any goal, no matter how challenging, as long as he still believed it possible. Ben Verwaayen, former co–chief operating officer, recounts the first time he had to inform his new boss (whom he describes as "extremely outspoken, extremely impatient, and relentless in his quest for results. Verbally . . . sharp as a knife") about losing a small piece of business. Verwaayen says, "It was like somebody had hit him in the stomach." Verwaayen tried to reassure McGinn. "You win some, you lose some," he told the CEO. But McGinn would not be consoled. "I don't want anyone here to be a good loser," he snapped. "We are winners."

Amidst the roar of Lucent's fans, Lehman Brothers analyst Levy now had some hard evidence to explain his nagging doubts. Lucent had missed its number by almost $1 billion, and some of its excuses troubled him greatly. Orders that could not be booked, Levy posited, were not simply a technicality: "Management believes that most of this revenue in the first quarter should be recognized in subsequent quarters, but our experience on Wall Street has shown that revenues that slip out of one quarter often slip altogether . . ." But it was the balance sheet where he saw the greatest cause for concern. In his report entitled "Lucid Views on Lucent," Levy peeled back layers of financial nuance to reveal a different company from the one his industry colleagues were still heavily touting. Lucent was not collecting its payments fast enough. This problem, Levy speculated accurately, arose from the fact that the quarters were back-end loaded, meaning that in the final weeks and days of a quarter Lucent grabbed whatever business it could book and threw it into the reported revenue number. If this happened in a single quarter it would not be a problem, but what if it were to become a bad habit? Levy saw this as far more than a hiccup.

Lehman Brothers was not Lucent's investment banker. It had not been included in the IPO syndicate, but Lehman's bankers held out hope that the telecommunications giant might one day throw it some mergers and acquisitions or other business. On Wednesday, February 3, 1999, Levy sent a draft of his report to Lucent's investor relations office. He knew that the conclusions would be controversial, and he could ill afford even the smallest inaccuracy. He also knew that on Friday, Lehman bankers had a long-scheduled appointment with Lucent. "Let me know if there are any errors," he told the investor relations person, and at the same time he gave his bankers a draft copy. On Thursday, Lucent got back to Levy with two words: "No comments." And later that day they canceled the meeting for the following day with Lehman's investment bankers. The message was, There is no need to come down to Murray Hill. We would never do business with anyone who would do this to us. Levy had crossed the line in Lucent's eyes. Lucent executives who previously had been happy to chat with him about the company's products no longer returned his calls. If he tried to make contact, he was redirected to officials in investor relations. Weeks after the report Levy bumped into a former colleague and Lucent sales executive, Nina Aversano, at an industry conference. When she saw him Aversano shook her head and said, "Steve, we're going to prove you wrong."

Despite Levy's warnings, McGinn's credibility with Wall Street was at its apex, and members of his own team were converts. "Four years ago, the businesses that became Lucent were growing in single digits," remembered Russo. "We thought Rich was a little crazy when he said he wanted to grow the top line in double digits. But we did it four years in a row." Following in Schacht's footsteps was not an easy task, but McGinn's success was so overwhelming that *BusinessWeek* suggested employees at Lucent were going to start asking, "Henry who?"

Through a combination of its own aggressive forecasts and the mania that had engulfed the country's stock markets, Lucent had few options but to continue to pursue the uncompromising revenue-generation strategy that it had embarked upon. But 1999 was not 1998 redux. The external environment had become decidedly less hospitable, and the actions Lucent needed to take in order to reach its inflated goals became increasingly desperate. The demand for second and third phone lines for Internet access was slowing, and it would be only a matter of time before demand for Lucent's circuit switches would follow. For years, Lucent had defied its skeptics and maintained a profitable business selling these mammoth switches. Now the decline was at hand. McGinn indicated that Lucent was looking to the CLECs to make up some of this deficit, but many CLECs were just beginning to experience their own financial ailments (which in most cases would prove fatal), and the risks inherent in doing business with them were rising.

Lucent had hoped to replace some of the declining circuit-switching business by storming the optical market, but here it had fallen far behind Nortel. Still, McGinn remained confident and refused to acknowledge the disappointing quarter as anything other than an aberration. His was a diversified business. As one door closed, another would open. Growth would not be a problem, McGinn insisted. Lucent was operating in a $400 billion marketplace and had only 10 percent of the market. McGinn never tired of telling his employees that the only thing they were not constrained by was opportunity. In many ways he was correct. The prospects that lay before the company looked limitless. There was universal agreement among industry observers and Bell Labs' own economic research that robust growth in telecommunications would continue at least until 2002. So what was the problem with a single bad quarter?

In light of McGinn's promises to his shareholders, the growth prospects for the company's business communications systems were troubling. Selling telephone systems to large corporations was a mature business with limited upside potential. This $8 billion business acted like a ball and chain on the company's overall earnings growth. The BCS business had become an annuity; the servicing of established equipment would continue to generate income, but, given Lucent's revenue targets, it was a drag on overall sales growth. In contrast to businesses like optical fiber that were growing at 60 percent, BCS was struggling with a growth rate below 5 percent. Each quarter McGinn voiced his disappointment in the department's performance, with the caveat that there was "no suggestion of selling BCS but we will improve it or move to plan B." Plan B was never detailed, but the assumption was that if spinning off BCS were an option, Lucent would willingly rid itself of this business. (The terms of Lucent's separation from AT&T stipulated that the distribution of Lucent stock to AT&T shareholders would cease to be tax-free if Lucent did any spin-offs of more than 10 percent of the company's assets for the first three years.) Management wanted to concentrate solely on those businesses where it believed growth could be accelerated. Anything that could not keep the pace, even if it threw off a steady flow of cash, was superfluous to requirements.

By 1999 Lucent was bleeding talent. In the midst of its race for growth, the company was shedding valued staff at an alarming rate. Over an eighteen-month period, William Brinkman, Bell Labs' vice president of research, lost 15 percent of his team to start-ups. He remembers that "there was a period of time where every morning someone was threatening to leave their job." Nationally, boom-time unemployment levels had dipped below 4 percent, to rates that most of those in the workforce had not seen in their lifetimes, and demand for those trained in the telecommunications industry far outstripped the number of job seekers. Lucent possessed two very valuable commodities, high-quality engineering talent and experienced marketing executives, and it was having trouble holding on to both. For companies hoping to steal market share from Lucent, particularly in the Baby Bell market, stealing employees was a reasonable first step. "It's like a jailbreak," explained Dan Smith, CEO of Sycamore Networks, about his successful efforts to hire people away from Lucent. Top management estimates that 25,000 employees walked out the door in 1999, and they were very difficult to replace. The turnover rate was 20 percent in 1999, a time when Cisco's defection rate lingered in the single digits.

All over New Jersey, small optical transport companies were springing up and siphoning off Lucent's personnel. Lucent had never spread its stock options as widely or deeply as its competitors in the start-up world. Job security, the prospects for advancement, and a pleasant work environment were the expected compensations coming out of AT&T. Now, suddenly, the rules had changed. As the social compact between employer and employee was being rewritten, Lucent's offerings were not enough. The effects were devastating. "Customers . . . are increasingly dissatisfied with our performance," Schacht would later say. "There is a perceived lack of sustained attention. A person that I met with just a couple of days ago reminded me that we had six senior VPs on their account in less than the last three years, and three account executives in the last nine months."

The turmoil was both distracting and debilitating. The company was reorganized for the second time in a year in July, from eleven businesses back down to four, creating further displacement. McGinn's hot little businesses had not worked out. While in theory the structure might have given Lucent a more entrepreneurial focus, instead it created confusion, replication, and excess cost. Each business established a number of staff functions that were performed eleven times over, with costs much more difficult to control. Different businesses sold to the same customers, sometimes offering overlapping product lines and competing products. Customers complained of confusion, and salespeople argued that the structure hindered their selling ability by creating unnecessary complexity.

To combat the lack of coordination Fiorina had established the "BUFKANS," or businesses formerly known as Network Systems. Under her leadership, this was a group of business leaders within Lucent who came together to combat the confusion created by the eleven separate businesses. They met regularly to coordinate their various businesses, which, despite being pulled apart, still depended heavily upon one another for their success. Competition for sales between departments, rather than with competitors, confused customers even further.

Schacht was highly critical of McGinn's structure. He had spent his entire tenure at Lucent trying to inculcate the idea of "we," and in slicing up the company into smaller competitive pieces some of this had been lost. "Not surprisingly, they said, 'Well, if I'm a general manager, I need my own P&L' and 'I'm going to run this as if I owned it' and 'Get out of my way,'" Schacht said. "And so each had their own market organization. Then very quickly each had their own sales organization. They had their own plants. They had their own purchasing agent."

Rod Randall was one of many employees who began their careers with AT&T, in his case in Bell Labs, and went on to work for a start-up (in his case Ascend) only to find himself back at Lucent through an acquisition. As chief marketing officer, he saw that by slicing Lucent's business up into small pieces McGinn had created confusion among customers, who were on the receiving end of competing sales pitches from different Lucent teams. A Baby Bell executive told him that Nortel and Cisco had each presented him with a single solution but that Lucent had left him hopelessly confused with multiple offerings from an array of Lucent teams. "There were several softswitch efforts: some from Ascend, some from Lucent, some from Excel [a Lucent acquisition]," Randall said. "There was one from the research division, one from the next-gen product division, one from the switching division; there was one from an internal Lucent ventures division, one from the wireless division. Oh yeah, and there was one from the optical division, all different flavors."

The defections were coming at every level. Of the nineteen top executives pictured in the 1999 annual report, half would not stick around for the 2000 photo shoot. McGinn had long relied on Carly Fiorina's judgment and ability to motivate Lucent's large sales organization. Fiorina was so effective in part because she never shied away from a tough situation. Lance Boxer, former president of software and onetime CTO of MCI, remembers that after Lucent was spun off, Fiorina was sent to MCI to try to do business. It was a thankless task, but Fiorina refused to see it as anything but an opportunity. Though MCI did not want anything to do with AT&T or its castoffs, Fiorina simply ignored the long-standing hostility and forged ahead. For McGinn, she was a sounding board and a possible heir apparent. As head of Lucent's leading business, selling to the largest telephone service providers, she was responsible for more than 60 percent of the company's annual revenues. One consultant who worked with management explains, "Carly was the informal CEO of Lucent. The whole organization turned to her to deliver the quarter. She was the one who would figure out the gaps, figure out how to make the quarter." When those who worked with Fiorina at Lucent reminisce about her, they always mention her talent and in the next breath her ambition. "I remember [Fiorina's husband] Frank saying years ago, 'Here's the reality: I have a wife who is going to be the CEO of a major corporation someday,' " recalls PR head Kathy Fitzgerald. Fiorina had been only one of the heirs apparent, but less than two years into his tenure as CEO, McGinn had given no one the nod.

As soon as Fiorina's face lit up the October 12, 1998, cover of *Fortune* magazine below the headline "The Most Powerful Woman in American Business," recruiters beat a path to her door. But Fiorina ignored them, following Schacht's advice: "Look, you've got to make up your own mind. It's not disloyal to think about other alternatives. If we can't keep you fully challenged then shame on us. But you owe it to yourself—and to our company—not to get distracted by the wrong kinds of offers. If it becomes known that you're considering offers, the headhunters will descend on you and you'll never be able to shoo them away. Decide very clearly in your mind what kind of job offer you would pay attention to. Whatever your goal is, don't talk to people about anything less."

One evening in March 1999, long after most of her staff had departed, Fiorina did answer her own phone, and soon afterward she was in McGinn's office. When Fiorina went to McGinn and told him headhunter Jeff Christian had approached her, representing Hewlett-Packard in its search for a CEO, he was extremely concerned. She presented him with an option that in essence said, "Make me president and I will stay." The presidential position had been vacant since McGinn took over the CEO's job from Schacht, but there were too many contenders for the spot. McGinn considered giving Fiorina additional responsibilities, perhaps as much as half the company, but to give her the COO title would have upset too many senior executives and more than likely resulted in even more resignations. "Rich was a strategy person, often focused on the outside market," one board member says. "He needed a COO to change the organization to make it the company he envisioned." But it had taken McGinn almost three decades to attain a position of unquestioned power. He had been forced to endure an extended partnership with Schacht, and he was not going to yield authority easily.

McGinn's own success was working against him. Hewlett-Packard wanted Fiorina precisely because she was from Lucent and because it was impressed with what the newer company had been able to achieve. After Schacht's departure, along with McGinn, Fiorina had become the face of Lucent and the person most closely identified with that success. So when Fiorina came to McGinn in the summer of 1999 with the Hewlett-Packard offer in hand, he had no counteroffer. Although he could cobble together an inflated compensation package, the only equivalent job was his.

The effects of Fiorina's departure are still hotly debated among those who have labored at Lucent. Many feel that Fiorina was responsible for much of the change in culture among Lucent's sales force that would exacerbate the company's troubles. She was responsible for the sales force, and she had set the tone. They argue that had Fiorina remained at Murray Hill, she might have been forced to take the blame for some of the events that followed her departure. Others disagree, among them consultant Dan Plunkett, who worked closely with Fiorina during and after her tenure at Lucent. He says, "Given Carly's record of achievement, both at Lucent and since she became Chairman and CEO at HP, it seems obvious that Carly's departure was a huge loss for Lucent at exactly the wrong time. The market conditions would have turned sour for Lucent whether Carly left or not. However, Carly's ability to lead is extremely rare. When Carly left Lucent a lot of Lucent's capacity to win left with her."

McGinn used the exodus as an opportunity to infuse and promote new talent. He had long been of the opinion that Lucent needed fresh blood to energize it, and the high turnover, disruptive as it was, gave him the opportunity to make changes. To his credit, McGinn promoted a number of people into top executive positions who came to Lucent by way of acquisition: Jeong Kim from Yurie, Robert Barron from Chromatis, Curtis Sanford from Ascend, and John Drew from International

Network Services (INS) were a few examples. Gone were the old AT&T insularity and complacency. "It's about restlessness," said McGinn. "A corporation is the sum of the restlessness of its people." Little of this talent would remain at Lucent, and with the onset of the telecommunications downturn Lucent would set back the clock and come to depend more heavily on AT&T insiders in top management.

"I AM OVER THE LINE."

By July 1999, Lucent's stock price was breathing some rarefied air, and as it crossed the $60 mark, up tenfold from the opening price, McGinn went to his board to discuss his surprise at the company's lofty valuation. He was too keen and knowledgeable a student of history not to know what happened to high-P/E stocks that failed to meet expectations. When the inevitable industry downturn arrived, he knew the most inflated stocks would be the hardest hit. That summer McGinn simply conveyed his concern to the board. He did not suggest that a downturn in the industry was imminent, but in his mind questions hung in the balance. Could Lucent grow its sales by another 20 to 25 percent in 2000 and make good on its promises to the financial community? Did telecom still have that much growth left? Would the market keep expanding in a manner for which there was no historical precedent? The debate seized the boardroom. Everything the company did henceforth would be predicated on the assumptions made in these meetings. The risks of sticking with the growth strategy were huge, but the risks of backing off were deemed to be even greater.

The best research at the time suggested that the industry boom was far from over. As long as the CLECs could continue to borrow, they would continue to build. If Lucent's executives tried to reduce the company's growth trajectory, to suddenly bring its P/E back to earth, they would be devaluing the worth of their acquisition currency and vastly denigrating their competitive position, or so the thinking ran at the time. Cisco's P/E hovered above 200. Its ability to snap up the most innovative small organizations would dwarf Lucent's unless Lucent stayed the course. Although the board sensed that a mania had enveloped the industry and, moreover, did not believe that Lucent could grow at 20 percent indefinitely, it did not feel that it could suddenly slow the company's frantic pace and concede a race with its competitors that it believed it could win. A soft landing still seemed possible. Bell Labs' own research showed that the telecommunications boom would probably last into 2002. The board believed that valuation levels in the stock market would contract over the following two years to more reasonable levels, and also that Lucent's own earnings would rise significantly over this period. The net effect would be a much lower P/E ratio for Lucent, but one in line with those of its peers. From this lower valuation level, Lucent would be able to gently deflate expectations without significantly hurting the stock price.

The board discussed the issues in these terms and came out in support of management's growth strategy. If the board had any problem with McGinn's ambitious plans for growth, this was the moment to reexamine the strategy. If it had any concerns about his running of the company, this was the point at which to speak up. Instead, McGinn was given the go-ahead. Looking back, board members argue that it was the best decision that could have been made, given the evidence at the time. Schacht recalls the debate that took place over the summer of 1999: "At the highest levels of the company we literally debated, 'Can you grow a $40 billion company 20 percent in the years 2000 and 2001?' And we concluded we could go another round. And we were wrong. The questions that were raised were the correct ones, but we got ourselves convinced that we had one more leg to go."

In telecommunications, along with its high-tech cousins, investors saw an opportunity for almost limitless returns. Whether sophisticated venture capitalists bringing to market embryonic start-ups or ma-and-pa investors looking for excess returns in their IRA, few wanted to face the fact that the deluge of investment in the telecommunications industry was creating an unsteady state of overcapacity and feeding the already inflated stock market bubble. Federal Reserve chairman Alan Greenspan defines a burst bubble as a decline in stock market values of 40 percent or more over an abbreviated period of time. A less specific, but perhaps more illustrative, definition of a mania is the moment when normal psychological conditions are inverted and investors and executives fear missed potential profits more than actual losses. By 1998 the switch had been made, and even industry veterans who were the main beneficiaries of the investment binge puzzled over the turn it had taken. Joe Nacchio, erstwhile CEO of Qwest and senior AT&T executive, said, "It surprised the industry how willing people were to invest in curious, if not spurious, business plans."

Policy makers, analysts, and economists followed a dangerous yet well-trodden path when they announced that the Internet had given birth to an unprecedented economic era, unimaginatively dubbed "the new economy." The argument was uncomplicated: Incomparable productivity growth had been brought forth by the technological genius of the Internet. Growth at a rate previously believed to be unsustainable without setting fire to inflation appeared now to be possible. The economy had been unchained. Economic conditions once thought to be out of reach were now all but inevitable. This would be an era so transformed by technological innovations that its productivity levels had been elevated to a permanently higher plateau. The vagaries of the business cycle, which had tyrannized economies since the industrial revolution, had finally been tamed. Belief in the new economy, with its twin tenets of high productivity and low inflation, had become a national religion by the late 1990s, and Greenspan, who was viewed as one of its greatest adherents, stated, "I believe, at root, the remarkable generation of capital gains of recent years has resulted from the dramatic fall in inflation expectations and associated risk premiums, and broad advances in a wide variety of technologies that produced critical synergies in the 1990s."

At this point the Federal Reserve might have stepped in to raise rates, stifling speculation before stock prices spun out of control. But the Federal Reserve's charter does not entail targeting asset prices, so as the central bankers closely monitored the economy's growth and inflation, the markets took flight. Stock market manias result in crashes and, in their most sinister incarnation, the destruction of a healthy economy. The difficulty of recognizing and forestalling these potentially calamitous conditions had plagued policy makers and economists since the 1929 stock market crash. Yet, even with the considerable analytical powers of the Fed at his disposal, Greenspan would argue that the presence of a market bubble could only be debated during its lifetime. The question can be resolved only in hindsight; it is only in the rearview mirror as we are driving away that its existence can be confirmed.

Looking back from the vantage point of 2002, McGinn was sanguine about the market's fickle nature: ". . . the stock market rewarded what the stock market wanted to reward—and that varied enormously from company to company. There were companies with no revenue that were getting rewarded handsomely. Sometimes even the prospect of growth or the hint of growth might be recognition or reward—or a recovery from poor performance. So there are many—to me, it's sometimes—it's Talmudic in terms of how one can understand what gets rewarded but I can tell you that when you do not perform well, you're typically not rewarded very well." As one century gave way to the next, the virtuous cycle created by investment, growth, and more investment violently reversed itself.

Although in 1999 the overall telecommunications market still looked robust, the environment for selling to service providers had become vastly more competitive. Capital investment continued to rise in 1999 and 2000, but this in turn attracted droves of telephone and data networking service providers into an already crowded and highly competitive marketplace. More troubling for Lucent and the industry at large was the fact that although spending on capital investment in telecommunications had grown on average 26 percent a year since 1996, total revenue for the sector had risen by only 10.5 percent per year. Demand simply could not keep up with supply. The burst of investment in telecommunications had created an enormous problem of overcapacity, fiber-optic lines that were laid and left unutilized, switching and transport capabilities that far outstripped demand. Service providers lowered the prices they charged their customers in order to recoup whatever they could of this investment. It is estimated by industry analyst Telegeography Inc. that 97 percent of the fiber laid was never "lit," or activated, and thus the pricing for traffic over the other 3 percent was determined by the massive overhang in supply. Advances in technology further drove down costs as the process of transmitting data became increasingly efficient. At the same time, a fleet of new service providers arrived and pursued a strategy of cut-rate prices in order to attract new business. Soon there were too many service providers and not enough people willing to pay for their services, and a price war broke out. For example, an STM-1 phone line that could carry 576 phone calls at once between the United Kingdom and the United States cost $12 million in 1999 and $1.8 million in 2001. The numbers were incontrovertible; return on assets for the industry fell from 12.5 percent in 1996 to 8.5 percent in 2000. If investor returns did not improve, money would migrate to other industries, weaker telecom players would run out of funds, and the entire sector would go into decline. It was an untenable investment model that in only a few more months would begin to crumble. Thus, at the very moment that Lucent was laying its most ambitious plans for growth, the industry was staring over a precipice.

"Rich, there is a lot of confusion in the system," Chief Operating Officer Verwaayen warned McGinn in August 1999. "We may have lost revenue opportunities due to the large number of organizational changes, too many products through too many channels et cetera." While giving investors good news for the remainder of 1999, Lucent was struggling internally to achieve its revenue targets, and trouble was brewing for the first quarter of 2000. The fourth quarter of 1999 still looked tough, Verwaayen told McGinn in a confidential memo, but, as he quoted controller Jim Lusk, "one way or another, we will make it." Verwaayen, known to his colleagues as someone who specialized in telling the unvarnished truth, was already concerned about what he was hearing from his employees about the first fiscal quarter of 2000, and in fact the full year 2000. He was concerned enough to alert McGinn, who was vacationing, about impending problems. In two recent meetings Verwaayen had held with the BUFKANS, a yawning gap in revenue was becoming apparent. Those charged with developing and manufacturing products were suggesting that the shortfall in 2000 might be as high as $1 billion. Those on the marketing side were telling Verwaayen that the gap looked more like $5 billion.

A climate of "revenues at any cost" enveloped Lucent in 1999 as the guidance McGinn gave the market was steered upward. "The pervasive culture at Lucent was 'Do whatever it takes to do a sale.' The potential sale trumped all else," recalls one senior manager from finance. Top managers within the company have described the intense pressure that was brought to bear in order to reach these public goals and the relief

CEO Robert Allen cut the cord with Lucent, giving birth to a company that would soon eclipse his own AT&T. *(Corbis)*

The Lucent logo stirred up controversy, inside the company and out. *(© James Leynse/Corbis/SABA)*

Henry Schacht and Rich McGinn, Lucent's first and second CEO, in happier times. (© *The New York Times*)

Nortel's CEO John Roth was the competition from day one. (© *Reuters/Corbis*)

Carly Fiorina shepherded Lucent's IPO and led its sales effort, but when it became clear the CEO job would not be vacant soon, she decamped for the top spot at Hewlett Packard. (© Reuters/Corbis)

Fiorina guided Lucent's new edgier image. It began with the very first advertisements. (*Lucent Technologies Inc./Bell Labs*)

invented
dial tone.

(ALSO phone, transistor, laser, Telstar satellite,

fiber-optic cable, cellular)

Have won awards. (Nobel etc.)

Specialize in making things that make communications work.

Will do same for you.

The former systems and technology divisions of AT&T, plus

Bell Labs, now Lucent Technologies.

Lucent Technologies
Bell Labs Innovations
600 Mountain Avenue
Murray Hill, NJ 07974-0636
1-888-4-Lucent
We make the things that make communications work.

Lucent's first day of trading was a moment of triumph for the newly coalesced management team. *(Lucent Technologies Inc./Bell Labs)*

CEO Henry Schacht brought Lucent to Bell Labs in Murray Hill so that there would be no doubt about the central role he intended the research labs to play. (© *The New York Times*)

Nothing reflected more intellectual glory on Bell Labs than the invention of the transistor. (*Lucent Technologies Inc./Bell Labs*)

As the ink dried on the largest technology merger in history, CEO Rich McGinn did not know the difficulty Lucent would have in absorbing Mory Ejabat's more free-wheeling Ascend Communications. *(Corbis)*

The Lambda router—This Bell Labs innovation (seen here through the eye of a needle) sent light zinging off miniature mirrors, but it came to market just as the crash took hold. *(Lucent Technologies Inc./Bell Labs)*

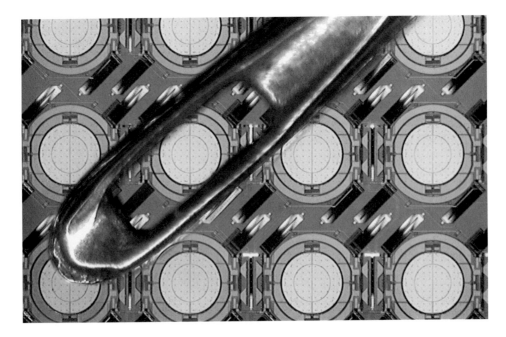

(above) Pat Russo ran Lucent's $6 billion business communications unit but was reorganized out of a job, only to return less than two years later as CEO. (© The New York Times)

(above, right) McGinn brought CFO Debby Hopkins in and was her staunch supporter, but many of those on the top management team were less enamored. (© Michele Asselin/Corbis)

Rich McGinn and Cisco's CEO John Chambers (right) discussed a possible tie-up between their companies, but soon the relationship deteriorated into acrimony. (© Alan Levenson/Corbis)

Lucent's Massachusetts plant, Merrimack Valley Works, was hailed as the cutting edge of its optical program in 2000, yet by the following year it had begun firing most of the nearly 6,000 employees. *(Getty Images)*

each time Lucent scraped through and "made the numbers." It was becoming clear that Lucent needed to take increasingly drastic action to reach its goals. Sales practices became decidedly more aggressive, and salespeople recall the atmosphere as the most pressurized in which they have ever operated. They sold software to customers for delivery a number of years hence, but much of this revenue was booked into the current quarter even though many of the software features had not yet been developed. This meant, in essence, that Lucent was selling products and capabilities that did not yet exist, while realizing the revenues in 1999. Schacht remembers the problem as "Nobody felt we were telling things we didn't think were true, it just turned out that what we told them we weren't able to do over and over again." Lucent "stuffed" its distribution channels with products that were often not resold, and although the revenue had been realized, the equipment was sometimes returned. An internal audit into Lucent's own sales practices suggested that its systems needed to be tightened. Lucent CFO Deborah Hopkins noted that such practices were taking place, saying, "I understand that we have cases where equipment was authorized to be sent to the customer without contracts being in place." She demanded a full accounting of the situations. Nina Aversano, a senior sales executive at Lucent who led the division engaged in these practices and found herself charged by the SEC with violating antifraud provisions of the federal securities laws, would later look back with some misgivings: "I would think if shareholders knew that we were pulling up business, gutting the future of the business in order to make a number in the quarter, then that would be a very disturbing piece of information."

Favored customers were offered gouging discounts on equipment they neither wanted nor needed at the time and for which they might have paid full price in later quarters. This practice became so common that customers soon refused to buy products on any other terms, pushing their purchases back to the final days of each quarter, when they knew Lucent's desperation would be the greatest. McGinn admitted that this practice was not only expensive, as Lucent sliced into its margins to make these deals, but also frustrated Lucent's abilities to predict the path of the business as sales were pulled forward from future quarters. During any given quarter it became more difficult for management to measure the company's performance, as the sales would all be back-loaded into the final week or two. The situation deteriorated quickly. In Schacht's words, "As we got further and further behind, we did more and more discounting."

In a concerted effort to realize revenues, each deal was scrutinized to find the last dollar that could be squeezed out and legally booked in the present quarter. The criteria were that title needed to have changed hands to the buyer, and that documentation and invoices were in place to prove the transaction was legitimate. Every quarter the pressure came down to the wire. "I would be reviewing contracts up until 11:59 P.M. at the end of every quarter," recalls controller James Lusk. "Even on New Year's Eve. I don't remember the last time I celebrated New Year's Eve." Sometimes in the final days and hours of a quarter, customers with whom Lucent had a particularly close relationship could be persuaded to take delivery of items they had not ordered. With their financing coming from Lucent, these customers were willing to provide assistance. For customers who shared warehouse space with Lucent, last-minute deals would be arranged where documentation was prepared and merchandise simply moved across the storeroom floor to the customer's side. Installation was even better for the books, and the push in the final days and hours of a quarter to get any system into operations was extreme. As salespeople told one another, "If it's carrying traffic, it's not coming home."

One very senior manager realized how far the company had gone astray when his boss telephoned him at an off-site meeting to say that a letter that he had sent to a supplier partner confirming a deal would need to be voided and the revenues, which had been tallied up for the following quarter but not yet recorded, would be erased. When this executive saw a copy of the letter, he realized the contents had been tampered with. He recalled, "Before I had left I told some of the people who worked for me to get a deal that was in the works done. I said, 'Do whatever it takes to get this done.' One person just went a bit too far." For a few individuals the pressure to tally revenues, which is acknowledged by most who worked at Lucent in these years as excruciating, was overwhelming, and they lost their way. The employee who had altered the deal documentation was immediately fired. The senior manager, a longtime AT&T/Lucent employee who had exhorted his employees to do their best, found himself taking a long, hard look at the new culture that had grown up around them. Without ever stopping to reflect, Lucent had slipped closer and closer to the line of acceptable and legal business practices, in his judgment. He gathered together everyone who reported to him and had a long discussion about what the words "Do everything possible to get this done" actually meant. He gave them strong words of advice: If you're worried, come to me, if we're worried, we will consult our lawyers; do not plow ahead when you have that niggling feeling that all is not well.

Months later, the same executive woke in the night with a startling thought: "I am over the line." Despite his lectures, despite being appalled and deeply concerned about his errant employee's behavior, he too was in the midst of negotiating a deal that slid uncomfortably close to the moral/legal divide. He recalls thinking, "I'm over the line, and now I was going to have to listen to my own advice and go to the lawyers and say, 'I think we are doing a deal that if it ever went to a discovery process, we would not want revealed.' " The climate had changed so much that "every quarter, everyone had a deal like this," one executive recalled.

THE QUALITY OF EARNINGS

There was no one sign on Lucent's balance sheet that indicated impending financial doom, but as the company closed the books on 1999 a series of symptoms suggested all was not well. While a single piece of financial information can easily be taken out of context, when a pattern occurs it is cause for grave concern. Looking back at Lucent's financial record in light of its later decline, one can see early warning signs of imminent financial distress in early 1999, when Wall Street was still blissfully in love with the telecom giant.

Revenues and earnings are an accounting manifestation, tabulated and recorded, as recent history has shown, in myriad ways. Lucent pointed to its income statement and elevated stock price as evidence of its strong performance. Revenues rose; the stock rose. It was a simple story; where was the problem? Many times, more accurate information can be gleaned by looking at the balance sheet. A few observant analysts who knew that in a healthy, expanding business, inventory and accounts receivable should not grow faster than sales picked up one of the early signs of Lucent's trouble. Lucent's focus remained one-dimensional, as Microelectronics chief John Dickson recalled: "Many times in . . . meetings I have heard executives ask for guidance on what they should concentrate on and several times I have heard the answer, 'just focus on revenues.' This is bizarre. No business manager should be exempt from simultaneously worrying about revenues, gross margin, operating margin, assets and productivity."

If the quality of a business is declining it will often show up as an increase in receivables, as customers struggle to make payments. It is a simple proposition. If accounts receivable is growing faster than sales, a company is not getting paid fast enough. Days sales outstanding (DSO) is a financial yardstick that compares the accounts receivables to the total sales for the quarter to gauge how quickly debts are being collected. Analysts like to look at this number to make apples-to-apples comparisons across or between industries. An increase in DSO arises either because the terms of sale are too generous and the customer is given too long to pay, or the seller has been too aggressive in recording a sale. A high DSO often indicates that a disproportionate and increasing number of deals are being finalized in the last days of the quarter or that the credit quality of the customers is declining. Receivables that are too long in the tooth end up as bad debt. For Lucent, the numbers looked weak. In September 1997 the DSOs were 84; a year later they were 92; by September 1999 they had risen to 108, soaring to 120 in 2000. Lucent attempted to explain away the increase in receivables by pointing to the increase in the amount of overseas business conducted, as foreign accounts often take longer to pay. However, the rate of growth in receivables outstripped the rate of growth of the company's international business, making this a less-than-compelling argument.

For every quarter in 1999, the growth in receivables outstripped sales growth, suggesting that Lucent's financial troubles were evident that year. The numbers would have looked even worse if Lucent had not engaged in the practice of selling its receivables. While on its face this might look like a good way to alleviate a growing problem, receivables are often sold at some discount to their face value (and in Lucent's case many were sold with "limited recourse," i.e., if the customer failed to pay, Lucent would be back on the hook), and thus the practice is a money-losing proposition.

When a company has rising receivables, the risk of nonpayment also rises. Many of those companies delaying payment are mired in their own difficulties and will never pay. Companies make an assessment of which portion of their receivables will result in nonpayment (an allowance for doubtful accounts), and then they charge that figure to income. It is therefore surprising that as Lucent's receivables rose, it chose to lower its allowance for bad debt, a counterintuitive and ultimately misleading change. One of the red flags that should have been waving wildly was Lucent's declining charge to income as a percentage of total receivables over the 1997–99 period. During this period, the assumption was made that the credit profile of customers was improving and that the rate of collections would be higher. Lucent had traditionally had a very steady rate of bad debts, at approximately 3½ percent, as its customers had fit a very similar credit profile. From 1997 until 2000, Lucent made a concerted effort to diversify its business away from the more stable Baby Bells and toward the CLECs and international customers. While under different market conditions this might have been a wise decision, under any circumstances the strategy increased the chances of bad debts. Even at the time, when it looked as though many CLECs would reach profitability, a shift in business from established telecoms to their fledgling cousins risked degrading the quality of Lucent's earnings. Under the most optimistic industry scenarios, the fact remained that the CLECs had much lower credit ratings and a substantially higher risk of default. When a growing segment of your customers is heavily indebted, loss-making CLECs, it is very difficult to see how your reserves for bad debt should decline. Reducing the allowance for bad debts improved Lucent's bottom line in the short term, but as expected, bad debts soon rose, and Lucent eventually had to take massive write-offs for nonpayments in 2001.

This is accounting information that Lucent makes public each and every quarter, but few observers thought it worth a mention. A buildup in inventory means that a company is manufacturing products that are no longer in demand by the market. Lucent's buildup in finished goods inventory rose 42 percent between 1998 and 1999, or more than twice the rate of sales, in what should have been a record-breaking sales environment. One of the arguments offered by management for this pileup of unwanted equipment was that component parts were in such demand and so difficult to obtain during this period that products could not be completed and shipped. However, inventories are reported as both total goods and finished goods, and the inventory buildup of finished goods was rising just as fast as that of total goods, so the problem was not just with parts. Some of this increase came from the precipitous drop in demand for legacy switching products. Demand for this obsolete gear would never return, and write-offs were soon taken. "We need serious teeth in inventory reduction programs," John Dickson of Microelectronics wrote McGinn in the summer of 1999. "Where are the hard and measured inventory goals? Who gets their rear severely kicked for missing target? The body language message to our organization is that the only serious interest we have is for revenue."

Dickson argued to McGinn that the problems stemmed from the company's distorted sales practices. He told his boss, "We have trained our sales force and our customers that special deals at the back-end of a quarter are the order of the day. No one can persuade me that our customers naturally take product in such a way that the third month has almost twice as much revenue as the sum of the first two. The stories of special bonuses for sales personnel are horrifying if only half true." The onslaught of orders that hit the company's manufacturing operations in the final weeks of each quarter as the salespeople desperately scraped together last-minute deals debilitated any planning process. The extraordinary became the ordinary as Lucent's manufacturing process lived in a world of "special expedites, just-in-case inventories, extra overtime and emergency shipments." Customers paid the cost, as Dickson recounted: "It also means that manufacturing personnel cut corners. Marginal product is released to customers. Incomplete systems are delivered to customers causing severe dissatisfaction as well as extra service and support costs. . . . Reject product bone piles are created for later resolution. Typically this stuff is never fixed and gets written down as obsolete inventory."

It is curious that at a time when product cycles were speeding up and Lucent acknowledged that it had missed the boat with the newest optical switch, the company lowered its assumptions for the percentage of inventory that would become obsolete and need to be written off. In all likelihood, 1999 was a year in which assumptions for obsolescence should have been increasing rather than decreasing, as the stockpile of circuit switches and dated optical switches looked increasingly less salable.

Any company involved in large, multiyear projects has on its balance sheet an asset called backlog. This is the portion of revenue that remains unrecognized in projects that are under way. For example, if Lucent received a contract from Verizon for $1 billion for work that was expected to take five years, some of the revenue would be recognized in the first year and the remainder would be booked as stages of the work were completed. Companies have a great deal of discretion in allocating the revenues over the period of a contract. Again, this backlog was a place where ominous signs could be found. Lucent's backlog went from $12.1 billion in 1997 to $10.2 billion in 1998 to $6.9 billion in 1999. This meant that one of two things was happening: either Lucent was getting fewer customer orders and the ones it held were winding down, or it was using overly aggressive assumptions and had raided this source of income. Either scenario was a cause for more red-flag waving. None of these accounting factors necessarily suggests impending doom, but each acts as an early warning signal, inviting investors to look more closely.

BANKING ON TELECOM

As a last resort, as the pressure to increase revenues rose and the revenues were simply not to be had, Lucent could always buy the business itself. Vendor financing, the widely practiced art of lending customers money to buy one's own equipment, is the murky backwater of the telecom industry. This type of financing occurs when an equipment manufacturer uses its investment-grade balance sheet to help its customers buy its equipment. It has been compared to seeking a car loan from the dealer, but the analogy is not completely accurate. Vendor financing loans are in effect steep discounts on the price of the goods because, unlike a car loan or home mortgage, in which the assets can be repossessed and sold, the resale value of quickly outdated telecom equipment is a fraction of its original value. The analogy also breaks down in cases where the service provider may already have offered the equipment as collateral to its bondholders; then vendor financing becomes like a car loan with no car. Vendor financing was one method by which a sale could be concluded quickly and the revenues booked into the current quarter. It helped make possible growth rates that did not seem possible. And vendor financing opened doors that were shut for a reason.

That Lucent was going to dip into this cesspool was no surprise, but few imagined the depth of its immersion. Lucent had offered such financing even as part of AT&T, and plenty of warning was given in the IPO document that this practice would continue. Controller Jim Lusk was an early skeptic. Lusk was a longtime AT&T/Lucent finance person and thought Lucent should be in the business of selling equipment rather than lending money. "I am old-fashioned about finance," he said, "I need to see when we are going to actually get paid back. I want to see what is going to have to happen, when it is going to happen, and if you can't prove that, forget it."

It was Lucent's first deal with Sprint PCS in 1996 that got the vendor financing ball rolling. Baby Bells did not need to borrow from their suppliers; they were sitting on investment-grade balance sheets. It was the new guys on the block that would come looking for money. At the time of the spin-off in 1996, Sprint PCS was looking for vendors for a massive wireless network build-out. In the request for proposals, Sprint made it clear that financing from the equipment supplier would be part of the deal. "This just didn't make business sense to me, lending them money to buy our equipment. I was worried," Lusk recalls. "But Sprint PCS was a grand slam, business with great margins." Lusk came around to the idea of vendor financing but maintained a healthy concern: "How much lending risk you are willing to take is really a business judgment, but as a finance person you know that somewhere something is going to go wrong." Lucent eventually won 60 percent of the contract, with Nortel selected to provide the remainder of the equipment. It was a spectacular win, as AT&T had never done any business with Sprint, WorldCom, or MCI. In the agreement Lucent provided $1.8 billion in financing, with all payment of principal deferred for four years.

The early lending successes were later used as justification for the practice—vendor financing was business as usual. But lending money to large cash-generative companies such as Sprint PCS (even one undergoing a massive network build-out) that already have millions of bill-paying customers is a far cry from making loans to brand-new CLECs still finding their way. One finance official who soon began to worry about Lucent's ability to judge credits in a changing industry asked, "The credit analysis was becoming very challenging. Sprint was a good deal, but as more and more people begin to offer wireless services and the competition becomes fierce, how do you measure the risk?"

All of Lucent's major competitors used vendor financing, although Lucent was particularly aggressive in its offerings. At the end of 2000, on the precipice of an industry collapse, the nine largest telecommunication equipment suppliers had $25.6 billion in loans to customers sitting on their books. Vendors defended the practice by pointing to the fact that only a small amount of sales were made this way and that, as all lenders say about themselves, they were very conservative and dealt only with the best credits. This was, of course, a fallacy, as most of the borrowers were credit-challenged entities that sought funding from their equipment supplier precisely because the more conservative banks had refused them. Afterward, when billions of dollars worth of these loans went bad, things would look different.

This was not an area that either investors or analysts chose to look at closely during the good years, in part because the amount of information filed with the SEC was not extensive and the financings in the 1996–98 period were modest. Like so many things that start off as a good idea, the first billion or so is carefully monitored, but as the billions mount up the standards seem to slide. There are times that vendor financing is undoubtedly helpful, a foot in the door for a new client or even the straw that tips the balance when a service provider is debating between two similar offerings. But for salespeople it is a very dangerous drug, or, as one of Lucent's investors pleaded, "Let's hope they can stay away from the special sauce. It's the quick fix for sales." As pressure to sell came down on those struggling with accounts that no longer needed or could afford Lucent's high-priced gear, what easier way to make a sale than to offer to pay for it yourself? When a department is facing the quarter-end crunch and the numbers are not quite there, what easier way to fill in the blanks? As one former Lucent salesperson noted, "We always had the vendor financing in our favor—so anything was possible." In the endless string of press releases documenting successful sales, few mentioned whether Lucent had had to pay for the business itself.

For the most part, Lucent lent money to companies that in turn used that cash to purchase Lucent's equipment. In some cases, Lucent went above and beyond the call of duty, authorizing loans to be used to buy equipment from other vendors (generally, when it was something Lucent did not manufacture) and for everyday working capital. Vendor financing meant that Lucent did not have to worry about providing either the lowest-price or the highest-quality product, because the financing would help sell the deal. Vendor financing was an ace in the hole, and its availability misled Lucent about its ability to compete based on the merits of its products. Lucent's leveraged balance sheet, rather than its success in selling products, may have created an inflated sense of the demand for its equipment.

Vendor financing exploded, as one might have predicted, when investment in the industry by other financial players slowed and the newer communications companies were gasping for cash. CLECs were the most avid users of vendor financing because many of their credit profiles were weak enough to place other forms of finance out of bounds. CLECs had long availed themselves of the high-yield, or junk, credit markets. "It was high-yield heroin," said Royce Holland, former president of MFS, an early CLEC. "You didn't need to do a road show with investors. You just had a conference call and you could get a few hundred million bucks." In 1998, the average high-yield telecom bond paid 8.9 percent and the total debt was $70 billion. By the end of 2000, yields were approaching 18 percent and the total debt was almost $200 billion. Venture capitalists had poured enough money into the telecom sector, and now they, the banks, and the junk bond dealers were in retreat. The CLECs were forced to look elsewhere. Their need for cash was unabated, but their sources were fast disappearing. When financial markets retreat en masse from continued extension of credit, they do so because the perceived risks have become excessive. Lucent along with its competitors stepped into this breach and, in doing so, further inflated the telecom bubble and eroded its own credit standing. Phil Harvey, editor of the industry Web site Lightreading.com, said at the time, "But what's being done to shore up cash-strapped service providers has many wondering whether today's big equipment vendors are walking through a forest fire on wooden stilts."

"The financing became the go/no go on the deal. Service providers based their final decision on the financing arrangements the various bidders could offer," one finance official who has since left Lucent said. "The salespeople would argue that this was their competitive advantage." Vendor financing soon became a pitched battleground between the sales and finance staffs. The pressure on the finance staff was overwhelming, and their ability to reject a deal decreased. Salespeople would make an announcement internally that a deal was within their grasp if only the financing could be arranged. In some cases, lower-level treasury managers refused funding to shaky companies only to find that the salesperson had gone over their head to the treasurer or CFO and been granted the funds. Vendor financing became the deal breaker, and in a company thirsty for revenues, the cautious words of the treasury department could be all too easily ignored.

The urge to lend to poorly capitalized CLECs was driven by more than just an immediate thirst for revenues. The old AT&T adage "The footprint is our friend" prompted Lucent to aggressively establish itself as the vendor of choice with the new service providers. Traditionally, equipment and systems from different companies were incompatible. Therefore the first vendor with a contract was able to dominate a given network, not necessarily because of the superiority of its product but rather because it got there first. Although this began to change over time, equipment contracts still tied a service provider to its vendor for many years, thus increasing the value of establishing an early foothold—an effort often made successful with the advantages provided by vendor financing.

The CLECs had their own reasons for continuing to build long past the point of satiation. A more fully developed network increases the quality of service and improves the likelihood of a lucrative buyout by a larger service provider. CLECs were focused not on the problems of an entire industry but rather on their own struggle to survive. And some of the networks put into place were intended to be not ongoing businesses but speculative investments seeking a quick sale.

Lucent's loan commitments in 1996 were a mere $156 million, barely a rounding error for a company with a $22 billion balance sheet. By 1997, this figure had risen to $1.9 billion, a significant increase but a manageable figure in boom times. Cisco and Nortel also jumped into vendor financing with both feet, but on this count Lucent outdid the competition, with twice as much financing as Nortel and three times that of Cisco. As the stress on Lucent's revenues appeared in 1999 and the CLECs' thirst for new sources of capital grew, Lucent made commitments to lend in excess of $8 billion, although the amount of actual cash flowing out the door never approached this figure. At the point at which your loan portfolio rivals that of a small midwestern S&L (with a much riskier customer profile), it is hard to argue that there are not some strange resemblances to a bank. In a glaring example of how vendors had displaced traditional lenders, at one point Global Crossing owed Lucent more money ($130 million–plus) than it owed its main bank, J. P. Morgan Chase & Co. ($100 million). As the lending practices escalated, some in the industry questioned their wisdom. "We think there's a natural boundary between banks and networking equipment companies," says Scott Kriens, CEO of Juniper Networks. "Operating in both spaces doesn't give sustainable leverage as a business model. We do help in certain situations where we think the service providers are bringing powerful new business models to market. But in general, buying routers from banks is a bad idea and buying money from networking companies is a bad idea."

Lucent's management had other ideas and even when announcing weak financial results revealed plans to keep lending. CFO Hopkins said, "I think it's how we'll keep the velocity of that portfolio moving and how we must run it like a bank and really think about it from that perspective." ("If they want to be banks," analyst Levy noted in response, "we'll value them as banks.") Hopkins tried to continue Lucent's lending with more stringent guidelines in place. She attempted to rid Lucent of the ad hoc lending processes and close off the back channels that salespeople used to gain approval, putting into place a vendor financing approval committee that would study each proposal for funding.

The notion of growing revenues by lending to increasingly risky customers was the most visible manifestation of a market that had lost its way. Lucent continued to insist that this was a carefully monitored business practice right up until the moment it was forced to begin writing off the bad loans. In fiscal 2000 Lucent took a $501 million charge to earnings, or 41 percent of the total earnings that year, because of these unpaid debts. While the company defended the soundness of its lending practices, Schacht would later brag that the decrease in financing commitments beginning in 2001 suggested a more disciplined approach to vendor financing.

By continuing to lend at such a rapid pace, Lucent and the other industry vendors were actually decreasing the chance that they would ever see their money again. They were facilitating the build-out of networks for which there would be no demand and that might very well not have been built without their help. In helping to create a situation of extreme overcapacity, they reduced the chances that their borrowers and in fact the industry as a whole would be financially viable. Too many competitors meant that most CLECs could not begin to garner the number of subscribers they would need to hit their break-even points. Commenting on the copycat plans of many of the CLECs, analyst Levy asked the questions that Lucent and its competitors should have been asking: "Do we need a 28th [digital subscriber line] network provider in Brooklyn? Do we need the sixth [third-generation wireless] network in Germany? In their drive to grow faster and faster, [Lucent and Nortel] may be artificially stimulating the growth of the market. And maybe some of these networks don't need to be built."

By pouring fuel on the fire in the form of easy cash, the telecommunications equipment vendors aided in the collapse of their own industry. As the situation deteriorated, it was not in the vendors' interest to take the road of fiscal rectitude and discontinue lending to struggling telecoms. Rather, now that they were bankers, they found themselves mired in John Maynard Keynes's classic banker's dilemma: "If you owe your banker a thousand pounds, you are at his mercy. If you owe your banker a million pounds, he is at your mercy."

Lucent and its competitors made vendor-financing loans to keep their own revenues flowing, but lending money in an immature industry, particularly when cash flows for the CLECs were projected to be negative for years to come, was an inherently risky proposition. Old loans ultimately need to be replaced by new loans, and when the cash flow to support them is not there, this sort of financing begins to look like a pyramid scheme. If the CLECs did not get more loans, they could not build their networks; if they did not build, they could not repay their loans. While in any individual case it might look as if the vendor is simply helping a customer over an uncomfortable bump in the road on its way to profitability, most of these transactions were prolonging the death throes of businesses that had become little more than sinkholes for cash. Lucent justified its lending practices with the age-old excuse that it lent only to the very best. By carefully studying the business plans of those companies to which it lent funds, Lucent believed that it was vastly reducing the risk of loss. However, no business plan projected borrowing costs in excess of 15 percent or a competitive environment crowded with droves of me-too CLECs. Many of these companies went bankrupt, Schacht argues, not because of faulty business plans or inept management, but rather because of a mistaken view of the available financing. Looking back after the telecom crash, Schacht believes that the one thing that should have rung bells all along the way was the fact that Lucent's customers, particularly the CLECs, were operating on the assumption that the flow of investment capital from both the private venture capital and public equity markets would stay open and cheap indefinitely. History suggests that this is not the case, and in hindsight Schacht can only wonder if a more thorough look at the past might have tempered the telecom spending free-for-all.

One of Lucent's largest and most controversial debtors was Winstar, a onetime ski apparel firm and erstwhile CLEC. Soon after Lucent's IPO, the young CLEC slid into Lucent's good graces by ordering twenty 5ESS switches. For Lucent this was great business, an opportunity to continue making money on its old standby product while at the same time getting a firm foothold in the CLEC market. The relationship was especially close, far more of a partnership than the typical customer-supplier arrangement. The fallout was personal, and the alliance ended in a high-profile $10 billion lawsuit filed in April 2001, with name-calling and recriminations that more closely resembled a celebrity divorce than a business deal gone wrong. What is clear is that each company benefited financially from the other, and that in at least one software financing agreement there were irregularities and misleading documentation that Lucent uncovered and reported to the SEC.

Winstar's technology, known oxymoronically as "fixed wireless," worked by setting up radio-wave receivers the size of dinner plates on the rooftops of major commercial buildings in large metropolitan areas. These dishes sent signals to other antennas in their line of sight and provided voice and data service at higher speeds and lower costs than traditional networks without the onerous expense of digging up the streets in highly congested areas. Their initial market was believed to be no less than every office building in America. But some of the technology had its drawbacks. It was affected by adverse atmospheric conditions such as blowing leaves, dirt, fog, snow, and rain.

In 1998, Lucent proudly announced that it had entered into a long-term strategic relationship to "build out Winstar's fixed wireless broadband telecommunications network in major domestic and international markets." Over a five-year period Lucent committed to providing the technology, network design, hardware, software, and money to establish Winstar's rooftop network. The $2 billion agreement (to be doled out in $500 million tranches) gave Winstar, in essence, Lucent's credit rating and access to capital and, perhaps of equal importance at that time, credibility. "It's quite a partnership. It validates our business model and expresses a huge amount of confidence in us," said Winstar CEO William Rouhana Jr. at the announcement. With the passage of time, it is hard not to ask what Winstar was providing in return.

The financing was unusually large and generous. Lucent lent Winstar funds above and beyond the cost of the equipment. With Lucent's support, Winstar's credibility rose, and between mid-1998 and 2000 its market capitalization tripled. Lucent had taken Winstar under its wing, and with its continued financial help, it was universally agreed in the halcyon days of 1998, Winstar would be able to complete its network and become profitable. Lucent serviced Winstar well, with a customer team of fifty people dedicated to covering the account, including program managers, systems engineers, and sales executives. In essence, Lucent was financing the construction of a network and exposing itself to all of the downside and none of the upside.

The infusion of cash from Lucent signaled to the market that Winstar was now more secure. Before Lucent's announcement Winstar's bonds were changing hands at sixty-five cents on the dollar (although they had been issued at par six months earlier), reflecting investors' concerns that the CLEC would run out of funds in late 1999 or early 2000. The day following the Lucent financing the bonds were trading in the mid-70s, pulling the debt of other CLECs upward as well. The perception was that new sources of capital had opened up for struggling telecom customers, making their success more likely and their debt less risky. Lucent was not simply aiding a customer in its time of need, but rather affecting the investment patterns and growth of its industry.

"Winstar is a focused, aggressive competitor who's positioned to win big," Fiorina said when the deal was struck in 1998. Lucent's top executives described Winstar's business plan as one of the most credible it had seen. Winstar would be a showcase for Lucent's next-generation networking technologies, she explained. Yet even as Lucent continued to pour money into Winstar, the CLEC gained unattractive distinction as the most heavily shorted midcap stock in America. Most CLECs would not be cash generating, nor did they claim they would, until many years after they were established. The number of customers they would be able to attract to pay for the huge capital outlay could be little more than speculation. The competitive landscape in telecommunications services was, as Schacht had predicted, entirely shrouded in fog.

Not everyone was thrilled with the new business. A number of managers from Lucent's finance department voiced their concerns about Winstar's financial viability and Lucent's exposure. Their voices went unheeded. One treasury official remembers, "People from treasury were coming to me and saying. 'This doesn't make sense, we shouldn't be doing this.' But each time I relayed their concerns to those above me the answer was always the same, 'This is incredible business and the risks are known and understood.' " Where Lucent salespeople were looking at a customer with growing equipment needs, the treasury department saw a borrower who burned through cash. Frustration grew among those within treasury, but they felt powerless to take on the sales juggernaut.

Later many would act surprised that CLECs ate cash for lunch, but in 1998 this fact was abundantly clear. Those lending to CLECs did not like to look at their borrower or the industry in such stark terms. The cold, hard numbers showed that Winstar's losses and debts were mounting, and in order to continue the illusion that this was not a highly risky credit, lenders needed an alternative metric that would highlight their progress. EBITDA, earnings before interest, taxes, and depreciation allowance, became the standard by which money-losing entities with huge debt and capital investment costs would be more flatteringly measured.

Throughout 2000, Winstar's losses kept pace with its rising revenues, a financial model without much promise. Despite this, it had little trouble finding new sources of funds. The company sold convertible preferred stock to Credit Suisse First Boston, Microsoft, and Compaq, payable in stock itself, which in turn diluted the equity owners' holdings. From an EBITDA position Winstar's finances looked to be improving, and by focusing on this deeply flawed measure of profitability investors were able to overlook its developing problems. The marketplace in which Winstar was operating had become extremely crowded. Over time the average account size had dwindled, and it had moved into smaller metropolitan areas. This had increased selling costs and the amount of debt needed to continue to operate and expand. Too many CLECs feeding at the same trough had produced a situation where there was not enough revenue for any.

By focusing on EBITDA and thereby excluding mounting interest payments, analysts said very little about the viability of a heavily indebted entity like Winstar, whose biggest problem was in fact its interest payments. It is quite possible to have $1 in EBITDA but owe $1 billion (and in this period, many companies found themselves in a similar situation). The one dollar in positive EBITDA provides very cold comfort for creditors. Even if Winstar's technology proved to be competitive in the marketplace (which was far from certain), the company had taken on so much debt that there was almost no possibility of it being able to meet its interest payments.

Even in March 2001, as Winstar's stock price descended from $65 to $2 a share, all but one of the analysts who covered the company for their Wall Street firms still rated it as a "buy," with an average price target of $40. Five weeks before Winstar's bankruptcy, SalomonSmithBarney reaffirmed its belief that the company was funded well into 2002. Its report took aim at a well-known short seller who had cast serious doubt on

Winstar's prospects, concluding, "It is evident that this short seller lacks an understanding of the CLEC industry. . . . WCII [Winstar] continued to remain one of the favorite names in the CLEC space, and we would be buyers at these levels—nearly their 52 week low." Shortly thereafter, Winstar's stock fell to fourteen cents. Winstar's assets, which the company had valued at more than $5 billion, were sold months later to IDT Corp. for a mere $42.5 million.

Caught in the dilemma that Keynes predicted, Lucent cut off Winstar's funding. Winstar could not make a $75 million interest payment and was forced to file for Chapter 11 bankruptcy protection. On April 18, 2001, Winstar went bankrupt, owing Lucent some $700 million. The mudslinging began in earnest the next day, when CEO Rouhana said, "Lucent's breach of its contract with Winstar has forced the company to seek protection under the U.S. Bankruptcy Code. Lucent represented that it had the expertise, personnel and financial wherewithal to undertake its obligations under the Supply Agreement. Little more than two years into the five-year agreement, Lucent has shown that its promises were hollow." The following day Winstar sued Lucent for $10 billion, arguing that Lucent had forced Winstar to break its covenants by not extending it more funds. Lucent argued that the responsibility lay with Winstar. "Indeed, Winstar is both in breach of their financial covenants and in payment default with us," a Lucent spokesman said. "We're clearly saddened by this latest development."

In the context of the late 1990s, selling equipment to CLECs made sense. The fact that the sector would implode was not known at the outset, and there was every reason to believe that small, technologically advanced service providers might come to dominate the industry. However, the line between selling equipment to the CLECs and lending them money was one that Lucent slipped over far too easily.

LOSING THE OPTICAL EDGE

Lucent might have muddled through 2000 if a little-known optical transmission system called OC-192 had not detonated like a bomb on bucolic Murray Hill. At another time or in another context, Lucent's decision not to pursue a particular optical system might have gone largely unnoticed. But at the speed Lucent was traveling, a car barreling down a highway at 120 miles per hour, one small steering mistake would send it off the road and into the trees.

Every technology company, if it lasts long enough, makes a product mistake. Some of these are legendary. IBM had its gargantuan mainframes, Sony had its Betamax, and Intel continued to make memories long after the market for its product was disappearing beneath it. Sometimes the mistake is an engineering one, a failure in product development that allows a company's competitors to speed by. Sometimes it is the failure to listen to customers or to understand competitors and the marketplace. Lucent's optical Waterloo would not come from traveling any of these well-trodden paths.

Bill Brinkman, head of research at Bell Labs, begged management for funding of the OC-192. He argued that it was the next generation of optical technology and that even if customers were not clamoring for it now, speed was the one underlying constant in the industry. He felt certain that when Lucent's sales force put the OC-192 in front of its customers, it would have no trouble making the sale.

Lucent's optical credentials were pristine. Alexander Graham Bell patented a device he called the "photophone" in 1880 that used light (including sunlight) and mirrors that vibrated with the pressure of sound waves created by speech to transmit sound without wires. Bell wrote to his father, describing his early wireless phone as his greatest invention, greater than the telephone. Thrilled with the possibilities, he wrote, "I have heard a ray of the sun laugh and cough and sing." But Bell could not find a reliable way to transmit light without interruption. The mystery of how to send voice traffic, and later data, over light waves was solved by Bell Labs and others with the invention of the laser, which provided a powerful, focused source of light, and the development of nearly pure glass known as optical fiber, which could carry light at, not surprisingly, the speed of light. By converting electrical pulses (electrons), the form in which speech had traveled across metallic telephone lines from the earliest days, to light pulses (photons), which slip down strands of glass at the standard 186,287.49 miles per second, scientists at Bell Labs revolutionized telecommunications.

By the 1990s, new optical products were needed to solve the two biggest bottlenecks that existed in networks: bandwidth (the amount of data that can run down the glass fiber) and speed. Dense wavelength division multiplexing (DWDM), which was developed in the late 1980s but came into widespread use only in the mid-1990s, made great strides in addressing the bandwidth problem. The system solves some of the congestion problems by breaking up beams of light into their color components and using each strand separately. With a powerful multiplexer, a one-lane highway of glass was instantly transformed into a 160-lane highway. The advantage of this technology was that it could be used to increase the capacity of existing fiber in the ground, even if it was not of the highest quality. Before Lucent separated from AT&T, Bell Labs had investigated ways of putting additional capacity into AT&T's fiber network. However, so much of its fiber was of an earlier, inferior standard that the optical carriers that increased speed were deemed unworkable, and this avenue of exploration was largely abandoned. Instead, Bell Labs focused on capacity, as few of the carriers had complained about speed and the projections for usage made capacity look like the impending problem.

Lucent became the number-one provider of DWDM systems. This was its chief optical transport system bet, and, in a major misjudgment about the direction of the industry, it decided not to attempt to improve the capacity of fiber-optic cable with a faster optical carrier. The best analogy for comparing the alternative technologies is a highway system where one product, DWDM, increases the number of lanes by splitting the light beams further and further while the second system, OC-192, simply increases the highway's speed limit. The technology decision boiled down to a bet on the rate at which the demand for bandwidth would grow.

Telecom carriers in the mid-1990s were satisfied with the industry standard that carried 2.5 billion bits of data per second over a strand of fiber. This was the judgment that Lucent management was heeding in 1997 when it decided to delay development of an OC-192 until 2001 or later. The OC-48 optical system would give service providers plenty of muscle to handle all the existing traffic flowing over their networks, but the problem of bandwidth congestion was looming. Some believed that rates in excess of this standard were not technologically possible; at faster speeds the lasers that transmitted the signal might melt the glass. But there was a difference of opinion on this issue, and both Lucent and Nortel had research teams exploring the possibilities of moving traffic four times faster (at ten gigabits per second) with an OC-192 system. Nortel's effort was aimed at a product that could be made commercially viable; Lucent's was a back-burner project. Brinkman was an advocate of the OC-192, and soon it would be the closest thing to a Cabbage Patch doll that the telecommunications industry had ever seen.

Lucent's management had decided not to fund two major development programs aimed at the same technologies solution and in 1997 opted to focus on DWDM over OC-192. By 1999, Lucent's optical business had become a huge success story, the fastest-growing product unit within the company. The 1997 decision looked to many like the right one. McGinn was confident about the course the business was taking and recalls, "In 1999, the optical business was our fastest growing unit—it grew nearly 40 percent—and the people in that business received the highest bonuses in the firm and accolades for their performance in the marketplace." At Lucent's analysts meeting in November 1999, McGinn remembers that "one of the highlights of that day was a discussion about our optical networking products, our architecture, our road map for evolution and the recognition that we were getting in the marketplace." No one at that meeting was shouting for OC-192, and, says McGinn, "That was not something that I was quite honestly concerned about in the first fiscal quarter of 2000." Brinkman continued to work on the OC-192 project, but with only the minimum level of funding. He believed that Lucent was underfunding this next wave in innovation but was unable to convince top management. As Lucent took the temperature of the analyst community for its optical plans and received the confirmation it had hoped for, Nortel charged off into realms unknown, marketing its OC-192 capability. But this analysts meeting led Lucent astray. Just because there was no barrage of questions about OC-192 (there would be in the ensuing quarters) does not mean that the analysts had not picked it up as a growing canker. Warburg Dillon Reed analyst Nikos Theodosopoulos pointed out days after the meeting that Lucent's optical business was experiencing a "drag on growth primarily due to a lack of OC-192 system and a declining market for OC-3." By

December, only weeks after McGinn had bathed in Wall Street's adulation, he knew that Lucent had made a major product blunder.

Nortel had not proceeded naively with its OC-192. CEO John Roth knew it was taking a sizable gamble. "There was a risk that our equipment would get too hot with lasers flashing ten billion times per second, or that the fiber in the ground wouldn't operate at ten gigs, or that the customers wouldn't want the capacity, or that the bandwidth demand wouldn't be as high as we thought," remembers Greg Mumford, Nortel's chief technology officer. But Roth believed that the demand for bandwidth was insatiable and ordered his team to fast-track the development of this unproven system. Ironically, a portion of the microelectronic components used to assemble the switch came from Lucent. Nortel would joke that it was able to get products and innovations from Bell Labs simply by ordering them from Lucent Microelectronics. Roth's engineers soon figured out how certain frequencies of light could work in a supercharged system without breaking down the glass, and Nortel began commercial development of its OC-192 product.

By 1997 Nortel had the product, but, just as Lucent had predicted, its customers were not clamoring for it—in fact, they did not want it. MCI tested Nortel's OC-192 system but walked away uninterested in the additional cost. Then, through a stroke of luck or genius, Roth made a deal that entirely altered the course of this technology. Early in 1997, he was trying to sell his still-unappealing system to Joe Nacchio, who was then CEO of Qwest. "Joe said to me, 'I have all this fiber and I need to light it up, so give me 2.5 gigabits a second,' " Roth recalls. "I said, 'That's not enough. You need ten.' " Nacchio declined the more powerful and expensive system. Roth then offered him the OC-192 system for the price of the OC-48 system he had intended to purchase. Nortel would get paid more only if Nacchio used the additional capacity that the more advanced system offered. Nacchio, not surprisingly, agreed. Within months, Qwest was paying for the new capacity, and the market never looked back. Qwest planned to build a nationwide fiber-optic network from scratch, an 18,500-mile road of glass crisscrossing the United States. "We wanted a pipe that would carry 10 gigabits without having to use a bunch of switches at OC-48 (2.5 gigabits) to bring it up to that speed," said Larry Seese, Qwest's executive vice president of network engineering.

When Nortel started this project it was the only manufacturer of OC-192. Lucent had led Nortel as the largest vendor of optical equipment. After the OC-192, that changed. Nortel went on to gain a 90 percent market share in OC-192, and its optical revenues surged from $4 billion in 1999 to $10 billion in 2000.

The situation for Lucent was dire, both financially and in terms of its image. "Your enemy has a machine gun and you have a pistol," a former Lucent salesperson said of selling optical gear before Lucent had a viable next-generation product. Once again, a company using Bell Labs innovations—and this time Lucent's microelectronic parts—had catapulted over Lucent. "They [Nortel] have gotten billions of dollars of business from a technology that we thought would be a niche business," McGinn later admitted. Optical was cutting edge; it was the fastest-growth business and had the highest profile. Claiming that an old business such as BCS, inherited from AT&T, had no future did not trouble the market, but failing to develop and deliver the fastest-selling product in the premier product category for the industry was a cataclysmic mistake. OC-192 reinforced all the worst stereotypes about Lucent, that even with the technology sitting in its laboratories, it would be late and unable to execute.

The postmortem was not pretty. Reviewing Lucent's troubles with OC-192 from the vantage point of 2001, Schacht said, "I've been into this one very carefully and our technology people absolutely were committed to the 10G [OC-192 optical switch]. They knew we could do it, we were ahead of the game, and we had it ready. If we had followed the recommendation of our technical people, we wouldn't have missed it. They were the ones who were saying, 'If you'd let us show this to the customers, they'd love it.' And in the interface between the customer and the technology people, we rejected that advice because we said, 'Look, we've tried it on the customer, and they don't want it.' The lab guys had it right, and we didn't listen to them."

Clayton Christensen's *Innovator's Dilemma* defines the situation Lucent was in, the pull between listening to the customer and seeing a world the customer has yet to envision. Lucent may have been listening to its customers, but as noted, Baby Bells do not live on the edge of technology; they are not where experimentation and innovation take place.

Their capital expenditure plans are set years in advance, and new products are integrated only after a painfully long testing period. That customers were not clamoring for a product that had not yet left the laboratory was not much of a surprise. In listening to its traditional customer base, rather than its more innovative competitors or its own researchers, Lucent had entirely misjudged the trend. Schacht proffered no excuses: "And shame on us—I mean us, big us! Not the people involved. All of us for not really realizing that there was a phase change available that we knew about but concluded the market wasn't ready for. That is a classic innovator's dilemma, and we just plain missed it! Unfortunately, our business history books are full of people who weren't able to solve that dilemma, and we've just written a new chapter."

The story was not quite over. Lucent's problems in its optical business extended beyond not having the right product to sell. As the product came on line, quality control issues emerged. In July 1999, Qwest's CEO, Nacchio, gave his former AT&T colleague McGinn an order for some optical equipment. Nortel had been Qwest's longtime supplier, but product shortages were acute and Nacchio needed additional sources. Lucent installed a system at Qwest that was riddled with technical problems and delays. The system had not been completely debugged, and Qwest was furious. Six months after the deployment, Qwest pulled the system and replaced it with equipment from Ciena. One employee from Lucent's North Andover, Massachusetts, manufacturing facility described a situation where quality problems at that facility led to product returns in April and May of 2000, "truck after truck after truck. Trailer truckloads in the plant. They had at least forty truckloads of returns. It sat out in the trailer, out in the parking lot. Millions of dollars worth of stuff just sat out there—millions! All OC-192."

In March 2000 at an optical fiber conference in Baltimore, Kathy Szelag, marketing vice president for Optical, faced down a group of anxious stock analysts. Her analyst audience was beginning to show the first signs of concern for Lucent's rosy projections. Most still had sky-high price targets for Lucent, and the previous quarter's weak financial news had not made anyone feel better. In 1999 demand for Lucent's optical gear was so overwhelming that production at its Merrimack Valley facility in North Andover was running twenty-four hours a day, seven days a week, shutting down operations only for Christmas and a state holiday. But while industrywide demand for optical gear rose by 14 percent in the first half of 2000 and Nortel's business expanded by 34 percent, Lucent experienced a 10 percent decline in optical sales. At the optical fiber conference, Szelag let slip that Lucent had put in place monumental targets for the rest of the year. One nervous analyst asked her, "How much of a stretch are the new optical targets?" Szelag replied, "These targets are a huge stretch and we have changed the way we develop our quotas now." She went on to explain that previously the sales quotas had been a negotiated item between management and the sales force; now they were set by taking into account the expected industry growth rate in each market. Nortel's optical business was on fire, so Lucent's targets were high. The same analyst, who had recently published a particularly robust forecast for Lucent, asked with a look of mild panic, "But Kathy, I know you said this is a stretch, but this number is achievable?" Szelag replied, "Achievable, but everything needs to go right." One analyst who observed this transaction remembers, "A few people were laughing but even they already had their cell phones up to their ears."

GAP FILLING, PULL-UPS, AND DISCOUNTS

At fiscal year-end 1999, Lucent looked triumphant. McGinn had kept his promise. He and his team had transformed Lucent into a $38.3 billion company, doubling its size in a mere four years. Lucent was the sixth-largest company in the United States by market capitalization. With 5.3 million shareholders, it had surpassed AT&T to become the most widely held stock in the world. One in ten stockholders tucked Lucent into his or her portfolio, and to most of them the company's prospects had never looked more promising. Lucent stock hit its all-time high of 59⅜ on December 9, 1999, an almost twelvefold increase from the IPO price and an astonishing return for its investors.

Yet as the year wound down and autumn gave way to winter, there were no illusions among management about the size or scale of the company's impending problems. The outlook for 2000 was bad. Nina Aversano and Pat Russo made an internal presentation on the 2000 operations plan. In it they discussed the increasing pressure to produce the double-digit growth that had been promised as Lucent came up against what they called a "revenue wall." The problem as they saw it stemmed from the fact that "legacy products alone no longer support that level of growth," and "Lucent's [market] share is lowest in highest growth areas (Packet Software, Access Services)."

One senior manager within Lucent explains that, although the company continued to tell an upbeat story to the financial community, it became increasingly clear by November that it would miss its numbers for the first fiscal quarter of 2000. In each of the two previous quarters, the situation had looked bleak and had been pulled through only in the final weeks; now, in the final months of 1999, it looked impossible. Another senior manager remembers that it was a "whispered truth" that the goals could not be met. No one was willing to stand up and point out publicly that the company's aspirations were beyond its grasp. No one wanted to be the department that let Lucent down, the executive who told McGinn that in spite of a booming industry, it could not be done.

Internal documents from mid-November show that while the analyst community was expecting $11 billion in revenues, and while McGinn had done nothing to dissuade them from this view, the numbers being fed from the bottom within Lucent totaled less than $10 billion. Another top manager recalls meetings during the fall in which a revenue "miss" on the order of $1 billion was openly discussed. Co-COO Verwaayen had warned McGinn of this possibility three months earlier. A private memo from CFO Don Peterson to other top executives dated November 3, 1999, reveals a dire situation. "As you are aware, we are faced with significant top and bottom line gaps in meeting our expectations for the First Fiscal Quarter [October–December 1999]," Peterson wrote to the company's officers. "The gaps we are facing will require a focused, combined and relentless effort to close over the next 60 days." Peterson begged for a renewal of faith: "Send me a positive confirmation that you are fully committed to driving your organizations to achieve our targets. Begin driving these actions now; don't wait for further clarification, be personally accountable." He offered some concrete solutions: "Stopping all management consulting work in queue, removing all temporary workers who have exceeded the 24-month limit, postponing Achievers

Club [the lavish weekends away for top salespeople], furlough non-employee workers for three days during Thanksgiving weekend—these are just some of the items that can produce quick results." Desperate times called for desperate measures, and even the smallest cost savings had to be examined. BCS, he told them, had limited travel to customer visits; Switching had restricted the issuance of cell phones to employees who had substantial business needs.

Yet McGinn still believed the $11 billion number was achievable. In the past things had looked bleak, but the sales force, under Fiorina, had pulled through. But Fiorina had left that summer, and he had little faith in the numbers he was getting because, as he was all too painfully aware, the gaming that had consumed the system meant that realistic number were no longer possible. Looking externally, McGinn saw a marketplace in which this level of sales was still possible, but inside the message was entirely unclear. He had heard all the excuses before—the doubts of his top executives were not news to him—but in the past they had always given Wall Street what they promised. So McGinn continued to assure the analyst community throughout the fall that Lucent would meet its expectations for the first fiscal quarter of 2000. And, as analysts passed this information on to investors, the stock price soared to its all-time high.

Investors may have been oblivious, but customers felt Lucent's desperation. To make their numbers, Lucent's salespeople scrambled to put together any deal that would fly. By calendar year-end 1999 one customer remembers that the typical discount, which had been 10 to 15 percent, had jumped to 20 to 25 percent. An even more damaging practice was the manner in which the salespeople robbed their future. "They offered discounts not only for that quarter but also on stuff that we were going to buy in 2000," one customer remembers. "That was a huge change. We typically had not received future discounts, only on the stuff we were going to buy right there and then." It was an enormous gamble, creating a crater that needed to be refilled every quarter. Schacht looked back upon these sales practices with misgivings: "There was a conscious decision made to meet the revenue growth, to try and meet the revenue growth by pushing and pushing, which resulted in increasingly heavy use of discounts to pull forward the profits of our existing business with the expectation that our new businesses would fill in the hole that was created. When the new business to fill the hole wasn't created, you darn well wished you hadn't discounted the way you had."

On January 2, 2000, McGinn told his board in a conference call that the quarter would be a disappointment and that the failure to make the numbers would come from everywhere. Lucent was particularly reliant on a few large customers, and this was where it would hurt. The AT&T account came up $250 million short, Saudi Arabia was short by $200 million, British Telecom by another $100 million, and on down the list of Lucent customers that were part of "deals that were on the table that fell off in December." While the message to the board was that these were done deals that at the last minute had fallen through, senior department managers recall discussions in October and November that the quarter would come up short.

With a few notable exceptions, every Wall Street analyst was still giving Lucent his highest buy rating. But to those inside the company or those outside who chose to scrutinize the myriad of SEC filings and market data, a different picture was forming. Inventories and accounts receivable continued to pile up. Lucent presented a puzzling picture, cash flow suggesting one thing and income growth another. Lucent's cash flow had turned negative in 1998 and accelerated rapidly to the downside. In a staggering turn of events, in 1999 Lucent lost $276 million on a cash flow basis while at the same time reporting an increase in earnings of 360 percent, to $4.8 billion. It was unusual for cash flow to continue negatively and earnings to continue to grow, and as the situation became more extreme it grew clear that something was going to give. Yet all of this might have been merely anecdotal evidence, and if the telecommunications market had maintained strong growth, it could have amounted to nothing but blips on the radar screen.

Lucent had a confident CEO with a stellar track record in an industry still on fire, yet there were a few Cassandras joining analyst Levy, suggesting that even as the company posted record earnings and analysts wrote flattering reports, the small print read quite differently. Over the four years since trivestiture, Lucent had made a series of legal accounting changes that enabled it to present its investors with a very rosy earnings picture. While no one had yet accused Lucent of illegality, accounting expert Brad Rexroad of the Center for Financial Research and Analysis commented, "It is pushing the rules." Most of these accounting alterations obscured Lucent's performance, making it much more difficult to accurately assess the health of the underlying business.

These were not accounting improprieties, and all of the rules changes had been disclosed in the company's filings and press releases and approved by its external auditors. Every decision was made within the "letter and intent of the law," explained controller James Lusk. But in a number of cases, accounting assumptions were employed that distorted the performance of Lucent as an ongoing entity and had the effect of making it appear a more prosperous business than it was.

The first adjustment to income came directly out of Lucent's separation from AT&T. In 1995, Lucent took a restructuring reserve of $2.8 billion as part of the cost of the spin-off. The reserve meant that Lucent posted a loss in 1995, and the funds were intended to pay for the severance of employees who would be laid off by the new company and for the cost of exiting businesses, such as the phone centers. By the following year, it was already clear that the reserve was larger than Lucent would need, and over the next four years, the company reversed this charge back into income. With discretion over when and how the income would be added back, Lucent was able to use these funds to smooth reported earnings until they ran out in 1999. In presentations to investors, Lucent made note of the fact that it was using these reserves but did not strip out this reserve reversal and present it as a one-off extraordinary gain. Rather, this windfall of $540 million was inserted into ongoing income from operations, thus giving investors an exaggerated impression of the underlying profitability of the business.

Another troubling accounting change came from two manipulations Lucent made to its pension fund. Ordinarily pensions are a cost for corporations, as they have to contribute to their employees' retirement costs. Lucent departed from AT&T with an overfunded pension plan. Under GAAP accounting rules Lucent was able to record some of this surplus as income. As the stock market rose, this pension surplus became greater and ever-larger amounts were taken into net income. In the second fiscal quarter of 1999, approximately 21 percent of the company's reported earnings came from pension income. The problem is twofold. First, although these funds may be recorded as income, they have little to do with the telecommunications business. While it may be argued that a rising stock market is suggestive of robust business conditions, this says nothing about a company's ability to provide products and services. Second, and equally troubling from a later vantage point, what goes up comes down. A bull market may have given Lucent an overfunded pension fund, but a bear market would do the opposite. Lucent was in the very best company when it boosted income in this manner, as dozens of other companies took similar adjustments, collectively misleading the public about the true state of their businesses. The pension funds of the S&P 500 companies were overfunded to the tune of $253 billion in 1998; by 2001 assets roughly equaled obligations, and in 2002 these same companies were facing an underfunding of $243 billion.

The second alteration Lucent made to its pension fund came when the company decided to revalue the fund with more aggressive assumptions about the expected rate of return. By assuming a higher rate of return on pension fund assets, the company was able to move even more money into income. This gave Lucent a onetime after-tax gain of $1.3 billion (noted as a onetime nonoperating gain) in the first fiscal quarter of 1999 and reflected the cumulative effect of making this change retroactively from 1986 to 1998. More important, this change in outlook allowed Lucent to lower its pension expense (or increase the amount of pension income) in the future by assuming a more cooperative investment climate. The problems with this accounting device are similar to those just noted. These adjustments reflected an optimistic scenario for a stock market at its peak but said little about Lucent's status as an ongoing business. Investors believed they were paying a lofty premium for earnings that came from building and selling technology. While Lucent never hid the fact that a portion of its income was from pensions—it could not, as SEC rules made such a disclosure necessary—the information was buried in the footnotes of the annual report and the SEC filings. As might be expected, when the market turned down, droves of companies began to reverse their optimistic assumptions about future pension gains and cash generated by the underlying businesses was redirected back into the pension funds. When Lucent announced a $4 billion restructuring charge in October 2002, it made clear that $3 billion of this was a charge to equity (rather than a cash contribution, which it still may need to make at a later date) because of the decline in the value of its pension assets.

The pooling-of-interest accounting that had helped make so many Lucent acquisitions financially viable was another spot where costs were hidden. In buying any high-tech company there is a certain value given to in-process research and development. This represents the R&D the acquired company has undertaken but that has not yet been translated into commercial value. If Lucent bought a factory, it would need to take the entire value of the buildings and equipment onto its balance sheet and then, year by year, depreciate the value of what it paid for, thereby reducing its income by the amount of the annual depreciation. High-tech companies hit pay dirt with a scheme that allowed them simply to write off the in-process R&D, although they paid for it in the acquisition price, and leave future earnings unharmed. This is permissible because the future value of the intangible R&D is highly uncertain. In this manner, Lucent was able to sidestep $2.5 billion in goodwill.

The sum total of these accounting manipulations was not fraud but a less-than-entirely-clear picture of how and where Lucent was making its money. At no point did Lucent seek to hide any of the financial information upon which these adjustments were made. However, little attention was drawn to the fact that a number of its businesses were turning south and others were receiving a big boost from liberal accounting maneuvers widely in practice across corporations at the time.

CHAPTER EIGHT   2000

2000 Stock price   High: $59.36   Low: $9.96

> *Think of this as a new season with new*
> *characters and a new story.*
>
> Kathy Fitzgerald
> Lucent vice president of public relations

> *A quarter of a trillion dollars lost in value.*
> *Someone should go to jail for what happened*
> *at Lucent.*
>
> Former Lucent senior manager

Entering the millennium, McGinn and his team knew they were facing an uphill battle, but no one could have predicted the disappointment this disastrous year would bring. It would be Lucent's annus horribilis, a year when careers and reputations were ruined, when those who had once been a close-knit team turned on one another with recriminations and lawsuits. Over the course of 2000, Lucent's market value collapsed by more than $150 billion from a peak of $258 billion. Putting that figure into perspective, in 2003 there were only eight publicly traded companies in the United States that had a total market capitalization of more than $150 billion. By the autumn of 2003, what remained of Lucent would be worth less than $10 billion.

Lucent's pain and turmoil began in the first days of the new year. On Thursday, January 6, 2000, after the stock markets had closed, McGinn faced the music. After months of rearguard action to piece together a quarter the financial markets would find palatable, Lucent announced its troubles to the world. This was not a regularly scheduled broadcast but an early warning flag sent up to say that, as the company tallied up the first quarter of fiscal 2000, the numbers looked bleak. The promised revenue growth of 18 to 20 percent would not appear; in fact, there would be no growth at all. Earnings had collapsed. No amount of pressure on salespeople, deals with customers, or shifting of profits from the pension fund could further delay the revelation that management's promises had been broken. To loyal shareholders who had only weeks before received further confirmation from management that Lucent's business was on track, the speed and magnitude of its reversal was a sudden and unpleasant shock.

Lucent had long basked in Wall Street's glory, but now it would feel its wrath. Management knew that despite its size and staid origins Lucent had been trading as a hyperactive growth stock, as hotly valued as many start-ups, and that any disappointments—given its P/E of 75— would crush the stock. At this valuation level, investors had priced in perfection, and the cost of failure would be astronomical. It was precisely the scenario that McGinn and the board had feared. Regardless of Lucent's lengthy track record of exceeding market expectations, the stock market showed no mercy.

Again, just as had happened in January a year earlier, McGinn took pains to explain his company's problems. "This is not a market issue," he insisted. "This is us not executing against a plan." The industry was fine; Lucent had failed to execute, and the problem was optical. The demand for every segment of the optical transmission process had soared—components, fiber, systems, and switches—and in each case Lucent had been unable to meet the demand or met it with the wrong product. In 1998 Lucent had held the number-one spot in the optical market, and by 1999 the company was running neck and neck with Nortel. By mid-2000 Nortel would hold a 43 percent market share while Lucent was struggling to hold on to its paltry 15 percent. In a market where service providers were desperate for the next optical advance, McGinn was forced to admit, "We gave our customers who had a need to meet capacity requirements a reason to go elsewhere."

Although they went unnamed, the two largest delays were from AT&T and the Kingdom of Saudi Arabia. Between these two entities, over $2 billion in sales hung in the balance. For many months McGinn did not know if AT&T was just playing hardball in its negotiations or if the business was not to be had. AT&T was moving business away from Lucent because Lucent did not have the right data products for the telephone company, but also because Lucent's success was ruining its relationship with its former parent and largest customer. As Lucent's stock rose twelvefold in its first three years, AT&T's stock merely doubled. As Lucent moved toward a more aggressive policy of distributing stock options, former AT&T colleagues seethed. "Jealousy developed that people who only a few years earlier were peers and equals now had differences in their personal wealth that ran in the millions," recalls one Lucent sales executive. "Resentment was such that pulling back their business was the only way they could get even," recalls another executive. "They were pretty angry that they weren't getting rich."

McGinn told investors and analysts of his disappointment but presented his case. He knew that Lucent had problems and had made mistakes in judgment. Yet, as he explained over and over, everything was fixable; product problems and customer delays could be quickly righted. While the quarter did not look good, McGinn stuck to his forecasts. Nortel's astounding and continued success in the optical field backed up his story. Demand in the communications industry was not weakening, the high-tech sector was not in trouble, and certainly the economy was still humming along. This hiccup meant that instead of exceeding the industry by 5 percent for the year, investors should now be looking for Lucent to best its competitors by 3 percent, still within the range of what had been promised. McGinn assured his investors that income growth would grow by 25 to 35 percent, right in line with what most of the rest of the industry was projecting. The problems, as unfortunate as they were, would be confined to the first quarter. The missing revenues would simply show up in later quarters, McGinn insisted. The credibility for his argument lay in the fact that just such a scenario had played out in 1999.

Levy was not buying it, although others were. In a matter of weeks,

Lucent's stock price had recovered most of its lost ground, yet Levy remained a skeptic about the future. While others saw a top-flight company operating in an exploding marketplace, Levy could not take his eyes off Lucent's balance sheet. Its accounting was too aggressive. "There's no fraud here," Levy said. "But this was a management that was trying to grow too fast. They dipped into the well too many times." Brad Rexroad, of the Center for Financial Research and Analysis, gave chapter and verse citing the pension fund, inventory, and accounting for bad debts in his detailed explanation of why Lucent was not really earning the numbers it reported. "They were trying to grow their business 20 percent. Their business just can't do that," Rexroad explained. Even after listening to all of management's arguments about the transient nature of their troubles, Levy remained troubled, saying, "Are the expectations the company is setting still too high, and can you reasonably expect another disappointment? I believe the answer is still yes."

Fifteen straight quarters of besting expectations did not soften the blow. On January 6, 2000, Lucent shares began the day at $72.75 (unadjusted), and in after-hours trading, where volumes are usually negligible, they ended at $52. The nearly 29 percent decline in value in a single day represented a $63.7 billion loss in market capitalization. Again to give perspective, in 2002 there were only thirty-five companies in the United States that had a market value in excess of $64 billion. No one within Lucent had fooled himself into thinking that investors and traders would be indifferent to the news, but the magnitude of the fall shocked all. Over the following twenty-four hours, 179 million shares of Lucent changed hands, an all-time record for any company and a number that would have been a good day's volume for the entire stock market only a few years earlier. The breadth of Lucent's stock ownership meant the pain was spread liberally.

Lucent was experiencing a very real downturn in its business. It was selling many of the wrong products and was struggling to ingest all of its newly acquired companies. But McGinn continued to believe (as all evidence suggested at the time) that the prospects for the industry were undiminished. He wanted plans for action from his staff rather than a litany of excuses for why his company could not perform. Part of McGinn's strength as a leader was in not accepting defeat, but his staff dreaded bringing him bad news. "He's almost physically ill if you bring him [bad] news . . . His skin turns gray," Lucent's COO, Ben Verwaayen, recalls. When presented with bad news by those who reported to him, McGinn insisted that they redouble their efforts or change their plan of attack rather than lower their expectations.

In February 2000, the economic expansion became the longest in history, and there were growing fears that the historic business cycle would rear its ugly head. Would the investment boom turn into a bust? Some investors grew nervous, concerned that the almost vertical rise in the NASDAQ could not continue. The NASDAQ had risen by 86 percent in 1999, and no one seriously suggested that such a performance, or anything resembling it, could be repeated. Some of the flagship names of the dot-com era were beginning to wobble. Gross domestic product growth in the fourth quarter of 1999 was 6.9 percent, a rate unanimously believed to be unsustainable. Business commentators and even the U.S. Senate began to focus on the question of how it would all end.

Stock valuations were so inflated by 2000 that the possibility of a gentle market drop and subsequent recovery looked to be quickly receding. Gradually deflating a stock market bubble is a challenge at any time, but there had never been a bubble like this one. There was no longer any painless way out. No longer would it simply be a few thousand dot-com millionaires getting the comeuppance the envious public hoped they would; now real people would get hurt. Between 1989 and 1998 the number of individuals holding stocks directly or through a mutual fund, excluding retirement accounts, had risen from 31 million to 49 million. If retirement accounts and contributed pension plans were included, fully 84 million Americans would feel the pain caused by a cascading stock market.

The parallels to the 1920s became frighteningly obvious: the new technology, low inflation, collapse in savings, surge of euphoria and investment, and abiding belief that the Federal Reserve had the situation in hand. The Internet was compared to railroads, automobiles, and the telephone in its life-changing force. Following each of these innovations, there was a dramatic increase in capital spending as the new technology was believed to give rise to a unique profit opportunity. The same pattern had recurred each time. Widespread optimism eventually incites the public to borrow or use its savings in order to buy more stock. As the public leaps into the unknown, investors can only guess at the potential return on these unprecedented opportunities, and in their enthusiasm will overestimate that which is possible. Many of these gains are later shown to be illusory, and as the laws of nature reassert themselves, sharply rising asset prices are followed by an equal and opposite reaction.

The Federal Reserve Board's responsibility includes "maintaining the stability of the financial system and containing systemic risk that may arise in the financial markets." The Board is charged with creating an economic environment in pursuit of full employment and stable prices, but former Federal Reserve chairman William McChesney Martin famously described the Fed's independent role as to "take away the punch bowl just when the party is getting started." Looking back, there had never been a party like the 1990s. Fed chairman Alan Greenspan had asked the now-famous question heard around the world, "But how do we know when irrational exuberance has unduly escalated asset values, which then become subject to unexpected and prolonged contractions as they have in Japan over the past decade?" The answer could be known only after the fact.

The one certainty in this increasingly insecure environment was telecommunications. No one doubted that this industry would continue to outpace the economy as a whole and that while capital spending might experience minor ups and downs, there were few more promising growth opportunities. Telecommunications permeated all other sectors of the economy, and improvements here would reverberate throughout the country. A fierce debate raged over how the industry and its many players and technologies would shake out—Would the Baby Bells fight back? Would Internet Protocol prevail over Asynchronous Transfer Mode?—but no one, absolutely no one, doubted the ongoing strength of the sector itself. Traffic on the Internet was believed to be doubling every one hundred days, or at a pace of 1,000 percent per year. While this was in fact never the case, this belief took on the patina of truth and was repeated by companies within the industry and even the U.S. Commerce Department. By extension, those companies involved in any way with transporting this deluge of information were thought to have limitless prospects.

Business investment doubled between 1990 and 1999. Under all conditions except one, this is unquestionably a good thing, a fact that was borne out by the commensurate increases in productivity and income over the decade. However, when investment turns into overinvestment, huge stores of excess capacity, be it in railroad lines or fiber-optic cable, are built. That eventually has the effect of depressing prices and profits for everyone in that industry. This is a particularly insidious and dangerous condition, and one that is immune to the Fed's traditional medicines. The usual mechanisms of stimulating growth do not work under these conditions. No reduction in interest rates, no matter how substantial, will induce investors to return to an industry already drowning in excess capacity and, by extension, insufficient demand.

Watching the enormous surge in technology investment, the Federal Reserve was forced into making a decision. Either the investment had led to a sustainable rise in the rate of productivity growth, thus altering the long-term growth rate for the economy, or it had led to overinvestment and overcapacity. Although the answer to this question would be known only in hindsight, the Federal Reserve needed to set monetary policy based on the expected outcome. Under the first condition the Federal Reserve could afford to keep interest rates low, as the capacity the economy was building would be used up in its accelerated rate of growth. In the second instance interest rates would need to be raised to rein in investment so that excess capacity, which would diminish the ability of entire industries to achieve an acceptable rate of return, would not continue to be constructed. The Fed believed that the investment boom of the 1990s was on the whole rational and that productive uses would be found for the new capacity created. In any given industry there might be some excess or shortfall, but on a pan-economy basis, investment was on course.

Greenspan believed in the power of technology to transform the economy. As he explained, "specifically the microprocessor, the computer, satellites and the joining of laser and fiber optic technologies" had created "an enormous capacity to capture, analyze and disseminate information. Indeed, it is the proliferation of information technology throughout the economy that makes the current period appear so different from preceding decades." Information technology had an additional kicker, making enterprises better able to plan and, according to Greenspan, "guard productive processes against the unknown and the unanticipated." With immediate access to data and improved control systems, a business could now monitor its inventory more closely using just-in-time inventory management. Many took this one step further and came to believe that these improved processes would help ameliorate the depths of an economic downturn by avoiding the deluge inventory that depressed prices and hampered profitability. On the final day of the century, *The Wall Street Journal* speculated that the "the business cycle—a creation of the Industrial Age—may well become an anachronism." Nothing could have been a clearer signal that the summit had been located.

Given the beliefs of the Federal Reserve, and Greenspan in particular, there seemed few reasons to rein in a benevolent investment boom. For telecommunications equipment manufacturers the Federal Reserve's policy decisions were particularly important, as they led to an overinvestment boom unparalleled in history. Part of the increase in bandwidth capacity was driven by innovations, such as faster switches and more efficient fiber. The industry's own calculations of demand could not keep pace with the ingenuity of its scientists. Rarely does an already deployed physical resource see an enormous leap in its productivity. Fiberoptic cable proved just such an exception. Before 1995, a single strand of fiber could carry 25,000 one-page e-mails per second. By 2002, the exact same strand of fiber that had done nothing but sit in the ground for seven years could now transmit 25 million of the same e-mails at no additional expense. At the same time that technology was expanding the existing capacity, increased investment meant that more and more fiber was being laid. This was not viewed as a problem because it was assumed that the demand for bandwidth was just short of infinite.

The equipment manufacturers were only too eager to expand to meet this projected demand. Like everyone else, Lucent's own research department continued to predict long-term growth for the sector. In the short term there were clear risks to expanding. Many customers were borrowing heavily to build out their networks, and if the Federal Reserve had misjudged the situation there was always the possibility of demand dropping sharply in the short term, as much a possibility as further expansion. However, while this was an acknowledged risk (discussed openly and regularly at the board level at Lucent), turning away real business because of vague fears of a blip in the economy did not seem to be a sensible decision. One and all, the telecom sector geared up for higher sales.

To stave off disaster, a CEO in this industry would have needed to predict accurately the growth rate of the industry, the eventual winners and losers, and the path of capital investment for the economy as a whole. Even if he called his own industry's developments perfectly, disaster would still unfold if the Federal Reserve made the wrong policy decisions. When the telecommunications industry hit the wall, the destruction that followed consumed all players. It was not within the power of any given CEO or company to control its fate. This does not mean that individual companies did not make poor managerial decisions and that the course of events could not have been altered somewhat. But the greatest damage to Lucent, first evident in 2000 and continuing through 2001 and beyond, was driven by macroeconomic and industry factors beyond its control.

Looking back at 2000, there was no hint of what was to come. At first, Lucent's difficulties looked to be its alone. Even as the company faltered, the telecommunications market looked to be steaming ahead. It was in this environment that McGinn continued to expect—and to demand—performance in line with the industry even as some of his employees began to tell him that they could not deliver it. A fundamental misreading of the macroeconomic environment led Lucent (and ultimately everyone else in the industry, albeit most of them to a lesser degree) up the wrong path, exacerbating many burgeoning difficulties.

BUZZWORD BINGO

Nina Aversano, attractive and energetic, came to AT&T in 1976 and was steadily promoted up through the sales organization of AT&T and then Lucent. She became the head of emerging market (CLEC) sales and later, with the departures of Russo in August 2000 and Fiorina in July 1999, she was made head of North American Service Provider Networks, the most powerful sales job at Lucent and in 2000 responsible for 78 percent of the company's sales.

Opinions on Aversano fall into two distinct camps. Analysts and many in her employ describe her as a capable manager, one who inspired her sales force and year after year delivered outstanding results. "It was okay with Nina if you took a risk," recalls Dawn Truax, a salesperson who worked for Aversano. "She was there to cover your backside. She wanted you out there swinging, because the one thing she did not tolerate was a lack of effort." Others who worked with Aversano, as well as Lucent's management, whom she later sued, describe her as a loose cannon who would spend weeks in a panic over sales targets, increasing the tension around her, and then triumphantly pull it all together at the last moment.

Uncontestable, however, is the fact that Aversano was a star at Lucent. Aversano was a natural-born deal maker who blazed a trail for Lucent into the CLEC market. In 1998, while she was still leading the emerging market providers sales force, she gathered her department at a sales conference at the Parsippany Hilton in New Jersey. For the three hundred sales managers this was a major event. Where else could you buttonhole the CEO for a chat or have lunch with the legal staff to iron out the nuances of a deal? Many emerging market salespeople were venturing into uncharted lands; only a year or two earlier they had been making sales for hundreds of thousands of dollars, yet now their deals were worth hundreds of millions and upward. Sales conferences were a place to swap stories, give and get advice, and hear from the company's top officers and scientists. Salespeople loved it. Truax remembers, "Nina persuaded everyone and made sure they were there. The legal, finance, and product people were all there, and salespeople loved it; especially if you were from outside New Jersey, it was a chance to grab these people's time." Management was grateful to Aversano's group and at this meeting announced a new compensation plan. As the hundreds of salespeople sat listening to their leaders, they were asked to take their right hands and reach under their seat. Taped to the bottom of each chair was cash, a token $5, $10, or $20, a gesture meant to drive home the purpose of the new compensation system. The message was simple: we will make you rich.

When Carly Fiorina took the stage with Aversano, her presentation focused on "advancing the ball." Fiorina was an advocate of decisive action, yet she knew as the deals grew larger and far more complex, uncertainty and hesitancy naturally crept in. She was here to soothe the sales force, to give them reassurance, and her message was resoundingly clear: make the best decision you can with the information you have and then act. Management, she told them, will back you up, and if the information changes, you can change course. Inaction was the demon. You are not going to get it right every time, she told her employees, but the industry was moving too quickly for anyone to seek perfection. Then she began to review the group's phenomenal results. Lucent continued to break its own records for sales and profits, and the emerging market sales force was at the forefront of that growth. Turning to Aversano next to her onstage, Fiorina asked, "What can I say?" And with a slow smile she said, "We owe it all to Nina." Then Fiorina, dressed as always in a pants suit, asked Aversano to place her hands behind her back and stick out her right toe. Fiorina got down on her knees and kissed Aversano's foot.

The push for sales subsumed all else and under Aversano this would lead to fraudulent conduct among some officers and executives, according to the SEC. Lucent was far from alone in this quest; as Arthur Levitt explained before the U.S. Senate in 2002, "The motivation to satisfy Wall Street earnings expectations was beginning to override long-established precepts of financial reporting and ethical restraint. A culture of *gamesmanship* over the numbers was not only emerging, but weaving itself into the fabric of accepted conduct."

Every Friday morning Aversano led a "gap closure call" to review the progress of her troops. Sales and product managers from all over the country logged on to a conference call where region by region across the country they listened to their colleagues give updates on revenue numbers for the previous week, focusing on what the total would be at quarter-end. The purpose of these calls was to develop the strategy for closing the gap between the orders in hand and the end-of-quarter targets. For the three or four weeks before the end of each quarter, and particularly at year-end, the meetings were held daily.

By 2000, these marathon meetings were starting two weeks into the quarter and eating up as much as 10 percent of a sales manager's time. As one manager remembers, "This would consume our day." Although designed as a mechanism for keeping management apprised and keeping pressure on salespeople, the exercise had long ceased to have its intended value but had become a game between the managers and the managed. These calls might include fifty to sixty people going over every detail of how each person would make his or her targets. Calls began at 8:00 A.M. Eastern Time and, one by one, over the course of up to five hours Aversano would grill her sales team on what they thought they could produce by quarter-end with the rest of the division forced to listen in. The game began to operate on many levels.

Customers who had long ago become acquainted with Lucent's extreme quarter-end pressure would not even dream of placing an order until the internal pressure had risen to the point where juicy deals were on offer. Paying full price early in the quarter was a mistake a new or foolish customer would make only once. With the ever-increasing pressure to book deals, the process became self-perpetuating. Anxiety rose among salespeople as the quarter proceeded, and this gave way to discounts. As the discounts grew, customers held out for more, and anxiety among the sales force and their management rose further still.

A sales manager who had some idea of a chunky order either in hand or as a reasonable prospect would never be so foolish as to mention it on Aversano's conference call until things really heated up near quarter-end. Naive sales managers who had reported good prospects or actual sales early in the quarter, in search of praise or recognition or out of simple honesty, had the experience of finding their managers in their offices raising their targets as the end of the period drew near. As one senior sales manager explained, on the weekly call it was his practice to tell 80 percent of the good news and 120 percent of the bad news up until the final days. This distortion of information, hardly unique to Lucent, made the task of reporting earnings and managing Wall Street's expectations even more difficult.

So the customers waited and the salespeople waited and the pressure to achieve the revenue targets promised by the CEO grew by the day. With many long hours needing to be filled on the weekly conference call, salespeople huddled around speakerphones playing round after round of Buzzword Bingo.

Born of desperate boredom, Buzzword Bingo was invented to provide amusement while listening to the unending excuses and gory details of another manager's quarterly sales. Buzzword Bingo cards were

created and distributed in advance of the call with a grid of the anticipated list of canned phrases that might be heard over the course of a
couple of hours. Each time a term was used the space was marked off,
with the goal of filling a whole row. The group would listen to hear
terms like "leverage," "step up," "pull through," or "stretch goals" as the
call dragged on. The winner then had to let the other players know of his
triumph by inserting "Bingo" into the actual conference call, perhaps by
waiting for a colleague to make a suggestion and then chiming in with
"Bingo, that is a good idea."

Aversano had studied under Fiorina and, according to those who
worked for them both, was her disciple in many ways. While at Lucent,
Fiorina had pushed her managers to set goals and then "stretch goals"
that they would publicly commit to trying to reach. This was one way in
which she had successfully inspired elevated targets. Aversano's methods were slightly different; she would ask for public commitments and
then seek to raise those expectations in private meetings. Salespeople
who worked for Aversano remember her entering their offices and closing the door behind her. Aversano would sit down and talk about how
thrilled she was going to be if the salesperson reached his target, but that
she had to push harder, the whole company had to push harder. She
needed a "personal pledge" from the sales manager, she would say—additional revenue above what they had agreed upon publicly. She would
then add, "You call me back and tell me how you are going to make this
additional revenue as a private deal between you and me." Some found
her intimidating in a one-on-one situation; others said they simply made
up numbers. Anyone who balked was not considered a team player.

Sales managers remember these days with more than a bit of resentment and misgiving. It is widely acknowledged that sales were made in
order to keep the stock price up, not because they were good for the customer or for the industry. "We went a little further than we should have,"
one senior sales manager remarked. Another noted, "The intensity of
competition got to the point where people started doing foolish things
for which the exit costs were very high." These were not simply sentiments born of hindsight. It became apparent to those managing sales at
Lucent that, over the course of 1999 and expanding in 2000, sales practices had become increasingly aggressive and in some cases question-

able. The SEC, in filing its case against nine Lucent managers and the company, said, "Moreover, on some occasions Aversano . . . affirmatively misrepresented facts to Lucent's CFO structure." Another sales manager explained that the hardest part of the job became passing the "red face test" with the internal auditors when trying to book disputed revenue. When sales executives are asked why this was allowed to continue and escalate, the answer is always the same: What could we do? The company was on a treadmill; it had to keep growing or die.

"The discussion focused only on the current quarter," recalls one of the most senior sales managers. "We never looked as far as four months out. The result of this was that visibility [financial projections] gets pushed off until the last possible moment. That's when the surprises come. There was always the belief that we would worry about today and that tomorrow would take care of itself. You cannot run a business that way." Another sales manager described the situation with a great deal of frustration: "Management thought that by creating an atmosphere of high expectations they were creating an atmosphere of excellence, but [it was] just the opposite, an atmosphere of mismanagement."

THE SLIDE

In April 2000, McGinn once again showed his determination to reinvent Lucent when he hired Deborah Hopkins as CFO. With anchorwoman-manicured good looks, Hopkins has an exuberant personality and enormous aspirations. "My mother loves to party, loves people," Hopkins said. "As children, we never had a box drawn around us like most kids do. We had no restrictions, not even curfews." A picture of Hopkins's mother riding an alligator graced her office wall. "Whenever I feel like I'm up to my ass in alligators, I look at that picture and think, Just go for it," Hopkins explained. Even before Hopkins came to Lucent, she was nicknamed "Hurricane Debby" for her aggressive turnaround efforts at other companies.

As a CFO, Hopkins had a solid reputation as a tough-minded cost-cutting financial expert, and McGinn believed she shared his view of a leaner, more dynamic company. When they referred to each other in public, they often used the term "partner." With this position, Hopkins would be ranked second on *Fortune*'s list of the most powerful women in America, after Fiorina. Her former boss, CEO Phil Condit of Boeing, said, "Debby could be the next Carly Fiorina." Hopkins had spent sixteen months as CFO of Boeing; before that she had held financial positions at GM, Unisys, and National Bank of Detroit. At Lucent, she began with a dramatic overhaul of the budgeting process. Instead of each department pleading its case, Hopkins and her team went about tearing the company's financial results apart, looking at each business and product line in terms of its profitability above and beyond the cost of capital. This budgeting practice, while hardly revolutionary, was new to Lucent and served to highlight those businesses that were not as profitable in the context of how much investment they required. Not everyone was thrilled with the change, but this sort of financial discipline is precisely why McGinn hired her.

Lured by what she described as the "passion and velocity of a high tech company," Hopkins failed in rule one of finance: due diligence. McGinn promised her they would be partners, and she, fearing that word of her negotiations might leak out, consulted with no one else at Lucent before joining the company. Hopkins faced more than a few surprises even before she had hung the pictures on the walls of her new office. In trying to keep her appointment quiet and her options open, Hopkins did not have the access to information about the company that she might have had in a more public job search. McGinn failed to consult the board of directors, a fact they did not take well. This may or may not have been Hopkins's fault, but in an organization genetically uncomfortable with outsiders it was not a good first move.

While she was at Boeing, Hopkins had made it publicly known that she coveted the top job, and the Boeing chairman made it clear that she was a contender for it. No one doubted that she had come to Lucent with similar ambitions. McGinn publicly stated that Hopkins would have a long runway at Lucent and that the CFO job "should give her a chance to be the CEO," to which she responded, "I definitely want that." Her ambition and confrontational style did not put him off. Even when she asked him what his timeline for staying was, he was not concerned because he saw her as a solid partner. Such flagrant ambitions at a company that considers people newcomers until they have been there at least a decade rubbed many who had spent a lifetime in service to AT&T and Lucent the wrong way.

Lucent had a language of its own, and salespeople working under Aversano had grown familiar with a variety of obscure terms. Over the years Aversano's sales team had evolved a vernacular to dissect its planning process that, even within Lucent, could be conflicting and unclear. Hopkins hated the vague Lucentspeak and made an all-out effort to eradicate terms like CTTP, or "closest to the pin," CWV, or "current working view," "aspirational view," "bottoms-up view," "stretch view," "50,000-foot view," "falling-off-a-log number," and "managing the funnel," all terms she believes were so amorphous as to be unintelligible. Then there were "Doug Fluties," the Lucent term for deals that were long shots, named after the Boston College quarterback who had thrown a "hail Mary" pass to beat Miami. Hopkins wanted a "forecast," a simple word she believed encapsulated the financial projections. She hammered at this point from the time she arrived at Lucent, making some progress but, in her view, not enough. The Lucent language had grown up over many decades, and try as she might, she was not going to quash it in a matter of months.

Hopkins could not have set foot in Murray Hill at a less auspicious time. Lucent's stock was trading at $47, but the momentum was not in her favor. While Hopkins began with a burst of enthusiasm, on April 24, 2000, she later admitted that she had been "handed a set of cards, and I'm making the most of the cards I've got. This is not the order I would have put them in."

To deal with some of its financial difficulties, McGinn announced in March 2000 that he was finally ridding Lucent of BCS in a spin-off to its investors. The spin-off, later to be named Avaya, was an $8 billion company. At the same time, the company announced that more manufacturing would be outsourced. Although this was a move its competitors had made many years earlier, it was met with acclaim. After the announcement the stock price rose toward pre–January 6 levels, as the market viewed Lucent as shedding one of the biggest drags on its growth. But the picture when compared to Nortel was stark. Lucent's stock was stuck at lower levels than it had been a year earlier, while Nortel's was almost 250 percent higher.

Relative Performance, Lucent v. Nortel, July 1999–July 2000

(01 July, 1999 = 100)

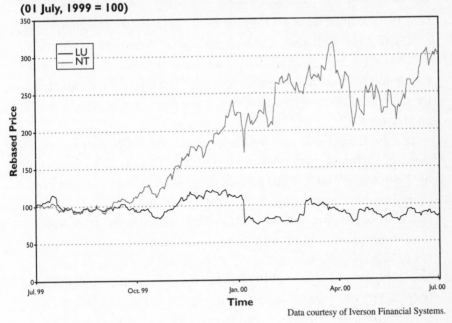

Data courtesy of Iverson Financial Systems.

The problem with the deteriorating stock price was manifold, and one that McGinn must always have dreaded. Lucent would be edged out of the acquisition game. The stock that had once looked like sterling currency had been devalued, eroding Lucent's ability to make acquisitions. Already priced into its stock was the boost the company would get in the future from acquiring faster-growing companies. If it stopped now, there was no telling what could happen.

New technologies would need to be homegrown, but that too was becoming a problem. Skilled engineers were far from immune to the lure of a rising stock price, and Lucent was having an increasingly difficult time bidding for expensive talent and, more painfully, holding on to the talent it had. "Our ultimate report card is the market," McGinn told his executive team. He knew that the stock price affected everything.

Sales in Aversano's North American business were falling behind. In April, she sent an e-mail to others in management saying that if "we were to continue 'business as usual' we stand to fall short of our North

America target by $200 million. . . . Simply put the business is in crisis and immediate and decisive action is required if we are to recover the year." She detailed what she believed to be the problems: "customer dissatisfaction (i.e. BellSouth, Cablevision, MCI)," "fractured accountability," "working at cross-purposes on the same account," "in addition to the obvious process flaws." She continued, "The inherent inflexibility of the model has fostered an unhealthy environment fraught with hostility and frustration."

In July, Lucent was once again forced to admit that earnings were slowing and that its performance, far from exceeding industry measures by 2 percent as McGinn had predicted in March, would not even keep pace. The third quarter came with many of the same old excuses and a few new twists. New pricing pressures had arisen in the marketplace, Lucent explained, and larger discounts were needed to win business. Optical deployment was still slow, although the larger story—that Lucent did not yet have an OC-192 product that was good enough to challenge Nortel's hegemony—was not fully revealed. Lucent's OC-192 had recently come on the market, but it was still riddled with installation problems, and the fallout from a series of technical mistakes, which had resulted from haste, was still raining down on the company. The Saudi contract was no longer "delayed"; Lucent was now "ramping down that project" and looking at $1 billion less in revenue from this customer. AT&T was going to come in $1 billion short of an expected $4 billion–plus in orders for the year. Lucent had no way of making up this $2 billion shortfall.

The long-heralded decline in circuit switching arrived with a vengeance. Lucent had long predicted the demise of traditional switch sales and in 2000 began to see confirmation of this dreaded event. Managing the cannibalization of your premier product line is a complex process, and despite knowing that it was inevitable, Lucent had timed it badly. For Lucent, this was not simply the phasing out of a technology but the decline of the company's preeminent business, the one responsible for more than 60 percent of its profits. Mercer Delta consultant David Nadler, who watched the process up close, explains, "Technological change, and especially technological substitution (where one technological solution displaces an existing one), often happens in a discontinuous and unpredictable manner. The problem is that this is the time when listening to the customer may not always be the best thing to do. Your customer will urge you to continue to provide and support the older technology, right up to the point at which they make the decision to substitute and move to a new technology. Suddenly, there was no more business. Technological change often occurs much later than people predict, but when it happens, it can occur with blinding speed." Revenues from this business plummeted from $7 billion in 1999 to under $2 billion in 2002. The revenue falloff had accelerated before the replacement products began earning their keep. "Next year, our wireless, our optical and our data networking businesses will each be larger than our circuit switching business," McGinn told his investors. "In the next two quarters, though, the year-to-year decline in circuit switching business will not be offset by top-line growth in other categories." These promises rang hollow to investors listening to a third round of bad news. McGinn continued to take responsibility and insisted that he took these disappointments personally, but that provided little comfort to investors, who were by now tiring of excuses. Fund managers had begun to dump Lucent shares after McGinn was forced to admit that the results were "not in keeping with the contract, quote unquote, we have with our investors."

"Part of what happened this quarter was not enough good old-fashioned discipline on cash collection," Hopkins said. She had now set up a "rigorous review process" to ensure that vendor financing was being made only to creditworthy CLECs (a phrase that would soon become an oxymoron), confirming for anyone who wondered that the previous process had been somewhat lax. The evidence of nonpayment was sitting on the balance sheet. For Hopkins, this was a troubling experience. She had walked into a mess that there was no way for an outsider to anticipate. She threw herself into the challenge, but perhaps her most embarrassing moment came after the July profit disappointment, the company's second during her tenure. Hopkins vowed, "It's not going to happen a third time. My credibility is the most precious asset I own."

"I don't have to tell you all how disappointing these results are for all of us," Russo had told the employees of her division on a conference call on January 26, 2000. "And in fact, what they've cost us in terms of credibility in the market. You read the same articles and analyst reports that I do, and so it's clear that this has dealt us a serious blow for sure but it's important that we recognize it's not fatal, but it is serious." Analysts who had been Lucent fans became cautious, and those who had been wary grew vociferously critical. "There is a good reason to be skeptical," said Levy a few months later. "They missed their December quarter numbers and lowered guidance for March. When they announced March, they lowered guidance for June. And now, when they announced June, they lowered guidance for September and December."

McGinn remained upbeat. Good times lay ahead. He and Hopkins had studied each of Lucent's businesses closely, and he confidently reported that revenues for the fourth quarter (ending September 30) would come in at 15 percent. To achieve this end, the company would undergo a strategic reorganization. The problems might be more of the same, but solutions would now be asset sales. Lucent announced that Microelectronics would join BCS as another Lucent spin-off. The reason given for slicing off this part of the business was that Microelectronics, later to be named Agere, had a strategic conflict with the larger Lucent. Microelectronics sold its products to all of Lucent's biggest competitors, with 75 percent of its sales coming from outside the company. Spinning off the division would completely eradicate this conflict and create, according to McGinn, two more agile, focused companies. Selling Avaya and Agere, the reasoning ran, would allow Lucent to move away from being the technology conglomerate that AT&T had bequeathed to become a more streamlined organization concentrating on the highest-growth areas of telecommunications. "We are creating a new Lucent that is totally focused on building the broadband and wireless Internet," McGinn said. These sales would begin the process of disassembling the company that AT&T had cobbled together when it spun off Lucent, eventually leaving Lucent as little more than the original Network Systems division and Bell Labs.

Putting aside strategic aims or distractions, Lucent had plenty of reasons to be optimistic about this spin-off. Microelectronics was one of Lucent's jewels, with more than $4 billion in revenue for the preceding year, and there was every reason to believe that given current market conditions and unconstrained by its conflict with Lucent it would perform even better in the near future. Lucent would be able to do what AT&T had done and shift some $2.5 billion in debt to Agere, despite the fact that the division accounted for only 11 percent of sales. This move would become all the more important as Lucent's credit rating declined over 2000–2001. The IPO might be a windfall for Lucent, potentially pouring billion of dollars in cash into the company.

McGinn sounded surprisingly like Allen five years earlier when he explained how shareholder value would be unlocked once Microelectronics was free from its "strategic conflict" with the larger corporation and Lucent would be free of the distraction of a noncore business. Both Lucent and Agere would be able to grow even faster, McGinn posited, with increased focus and a singular mission. Agere was in a growing market, and analysts estimated at the time of the announcement that the new company could be valued by the stock market at $100 billion or more, suggesting that, like Lucent, it might become the largest IPO ever.

HOLDING OUT HOPE

Microelectronics was such a hot property that McGinn soon found outside suitors were interested. In September 2000, Craig Barrett, Intel's CEO, called out of the blue, saying that he would be coming east and perhaps the two might meet. At their meeting in New Jersey, Barrett told McGinn that Intel was looking to move more aggressively into the telecommunications business. The optoelectronic components business was particularly enticing. Intel had recently spent more than $6 billion acquiring businesses in the communications industry, and he was interested in Lucent's Microelectronics. This was going to be a major strategic direction for Intel, and Lucent owned the number-one property. McGinn was all ears. Having already announced the IPO, McGinn would need to offer Lucent's shareholders a better deal than they would get in the marketplace if he was going to sell. But this little conversation might yield a mountain of cash and Intel stock. A team of executives was put together, with Bell Labs president Arun Netravalli and Lucent executive vice presidents Robert Holder and William O'Shea, along with Leslie Vadasz from Intel, to explore potential terms that would make the deal possible. But as Intel's own business showed softness, it began to second-guess the wisdom of making such a sizable investment. No terms emerged, and later in the fall, as the telecommunications industry faltered further, Intel lost interest. The repercussions of such a deal, had it been reached, would have been enormous. Intel might have suffered one of the largest write-downs in corporate history as the Lucent Microelectronics business collapsed. With its coffers full of Intel cash and Intel stock, which held its value far better than Lucent's own, Lucent would have been in an incomparably stronger position to weather the upcoming storm.

By spinning off Microelectronics (Agere) and BCS (Avaya), Lucent was now admitting that it would no longer lead the networking revolution with end-to-end solutions and best-in-class leadership in every product category. It was perhaps a more realistic approach, but coming as it did after three disappointing quarters, it signaled defeat. "The idea that we could be the leader in 18 different areas probably isn't the right way to think," said McGinn. But he added, "No one should doubt our ability to reenergize this business. We're sitting on a gold mine."

McGinn looked at the spin-offs as an opportunity to regroup and redesign the company. Onlookers were highly skeptical, suggesting that Lucent had lost its direction and was grasping for solutions. The company had frittered away its preeminence in the optical market, there had been turmoil and defections in many of its acquisitions, and despite targeting Cisco as enemy number one, Lucent had not been able to penetrate the enterprise market for data products in any substantial way. Cisco dominated this market and continued to report record results into 2001. Lucent's management insisted that the problems were quite simple and boiled down to nothing more than the OC-192, supply difficulties in other optical products, canceled contracts, and a falloff in its legacy businesses. It thought that the market would give it some breathing room, that the new initiatives would reveal a company in charge even if in the midst of change. It was wrong.

Lucent's continued insistence during the summer of 2000 that it would grow at or above the industry average flew in the face of the recent evidence about its performance. The switching business was down, two of its largest customers had retrenched, and although it continued to point to the growing market in optical products, Lucent's own business was contracting. In the second calendar quarter of 2000, Cisco, which had a much smaller foothold in this business, watched its optical system sales increase by 53 percent for the quarter, while the market leader, Nortel, saw its optical revenues climb by 34 percent in the same period. While Lucent still held the second-largest market share after Nortel, its business actually declined by 10 percent. Lucent had staked its reputation on its optical business. While the company pointed to a single product miscalculation and component shortages as the source of its troubles, the question that investors were asking themselves, in public and private forums, was: If Lucent cannot increase its sales in optical systems when it developed many of the technologies itself and the market is undergoing the greatest expansion it has ever witnessed, where can it?

The situation had, by July, become critical, and every level of management was exhorting the one below into action. Russo sent a message to Aversano and other managers, pleading for renewed attention to optical sales. She said, "We need to think of the issue with optical sales as a crisis. We must ramp up our coverage, customers, contacts and sales in this space. I ask you to work this issue as if it is the crisis it is. . . . We have got to change the momentum going in this arena immediately. I know there is a lot going on but I am concerned it is not enough to change the course." The conversations on the August sales call were simply a repeat of July with added tension. Lucent was now trading on hope, hoping to grow its way out of some of its difficulties. Its OC-192 had continuing problems with installation, yet there was hope (which was not to be gratified) that when the product became competitive, demand for it would still be raging. There also was hope that Baby Bell capital expansion plans for 2001 would be an improvement over 2000, and that the CLECs, which were facing a cash crunch, would find new funding sources (apart from Lucent's own coffers). Given the events of the previous five years, these hopes might not have been unrealistic. With the raging demand for products that Lucent's competition was still enjoying, there were reasons to believe, at least well into the summer of 2000, that better times lay ahead.

There was hope and there was confusion. Lucent had grown too fast, tried to incorporate too many new entities, and the strain on its systems was evident. "We were running on systems and processes that we inherited as part of AT&T. In growing the business so rapidly, the financial and information systems that we inherited were not able to keep up with the changes in our business," notes made for a Schacht interview with *Fortune* magazine read. The AT&T systems were so antiquated that "some processes were still being done manually to accommodate legacy systems," and as a result, the notes continued, "our visibility became clouded." Bill O'Shea, executive vice president for corporate strategy and marketing, admitted that Lucent could not accurately forecast product demand, revenue, or earnings because it had never fully integrated its own systems with those of its thirty-eight acquisitions. "Information doesn't really flow through them," he said, "so as a result we don't get a holistic view of the business until literally after we start closing the books. And of course, that is much too late to make the kind of predictions we need to make in order to give people appropriate guidance." As the telecommunications market imploded with a speed and severity that surprised all, Lucent's own information systems could not take the pressure, and its promises to Wall Street soon had little meaning.

It was not only the information systems that were clouding McGinn's view. The head games played between different levels of sales management made it difficult for him to get a clear picture, because it was deliberately being fogged over by those below. If salespeople in the field were leaving the good or bad news until the last moment, what chance did managers have of keeping investors apprised? Were the pessimistic midquarter forecasts an accurate reflection of deteriorating business conditions, or would a few salespeople triumphantly pull the aces from their sleeves at the last minute? McGinn had no way of knowing.

Lucent's overly aggressive growth strategy also hurt operations. By "pulling forward" sales that customers might otherwise have deferred, the company diminished its ability to forecast demand and violated accepted accounting procedures, thus misleading its investors, according to the SEC. Deborah Harris, a vice president in Lucent's customer team, had explained to Aversano in September that Winstar was "vehement that [it was] out of money and [did] not want to spend money on product that [it] cannot immediately utilize. The deals of the past are haunting us . . . there's $87 million [of inventory] in their warehouse." Schacht says that in 2000, one product development unit simply had to guess what customers would buy in 2001. Sure enough, the unit guessed wrong, but by the time this became clear, Lucent had manufactured a pile of products that were out of favor. Such problems explain in part why inventories grew by 34 percent in fiscal 2000, outpacing annual revenue growth. When asked whether she had been aware that in the final quarter of 2000 Lucent had depended heavily on "pull-ups," Hopkins offered up a blunt truth: "I'm aware that Lucent had been doing pull-ups for two years."

Surrounded by disappointing news, McGinn had come to depend on John Dickson, the president of Microelectronics, to make up for shortfalls in Lucent's other businesses. Dickson, a discreet Englishman, had come to Lucent/AT&T in 1993 from ICL, a British computer systems company. McGinn valued Dickson's managerial skills and described him as running his business with "insight and precision." Dickson was highly respected by his colleagues and acknowledged by insiders and observers as one of Lucent's most capable executives. Even as other businesses began to disappoint, Dickson continued to provide his boss with better-than-anticipated results. It is little surprise, then, that as Lucent's problems mounted, McGinn turned to Dickson, asking for his counsel and an honest appraisal of the company's woes. In his confidential missive, Dickson did not hold back. While he apologized in advance for anything that might sound arrogant, he presented McGinn with a scathing assessment of the business's problems and the radical solutions that would be needed to coax a world-class performance out of Lucent.

Dickson's "brain dump," which runs to three single-spaced typed pages, points to the failing performance of his colleagues: "My overall observation is that the business is out of control and operating 100% in tactical mode. As a general rule we don't have bad people in the business but we do have ineffective people and we also have a lot of effective people who have been trained or conditioned to do the wrong thing. . . . We also maintain on the payroll very senior managers who are seen by the vast mass of the organization as failures. This sends an awful message to the organization about our and your leadership skills and has a corrosive effect on morale especially when the organization is being pushed so hard each quarter." After pointing to the management of

Lucent's optical business as a major failing, Dickson asked McGinn, "You may know that they don't have a future but the great unwashed are not aware of this. Why not just send them home with pay if we want to be kind?

"We operate without a strategy," Dickson continued to tell McGinn. "Everyone and everything is consumed in making the quarter. No one thinks about the future so the fundamentals never get addressed and we never get off the treadmill." Dickson suggested that McGinn needed to raise the standards of performance and accountability among his management. It was a sentiment that executive vice president Bill O'Shea would echo strongly only a few weeks later. "We need sanctions in the organizations for non-performance," Dickson wrote. "You once said at a Lake Placid meeting that you didn't want to do this and it needed to be done elsewhere in the organization. This is fine but it is not being done anywhere."

Dickson finished with a dire warning for his boss: "Rich, the situation is extremely serious. There are far too many conversations amongst your close leadership team about this thing 'being over' or 'terminal.' Without change in management style these conversations could be well founded." Dickson did not hold back on his criticism of his boss: "My frank suggestion would be for you to get yourself a real president or CEO with real power to look after the business as Mr. Inside. And for you to do what it is you are one of the very best at, Mr. Outside. . . . With the right sort of focus and applications we could still have [Nortel CEO] Roth and [Cisco CEO] Chambers on the ropes. My apologies for rambling on so long. Hope this helps."

Advice to the CEO under duress came from all quarters. Bill O'Shea had worked for McGinn for a decade and a half and was brutally honest with his boss. Lucent was fast losing ground, O'Shea explained in an August 14 confidential 5:00 A.M. e-mail to McGinn: "In spite of considerable effort and success in several areas we are a company in *crisis*. We are growing significantly slower than our key competitors. We are losing ground in key growth areas for the future. We are missing our plans and losing the confidence of investors. And our people are disenchanted and losing confidence. . . . We must pull together and fix this fast." O'Shea told McGinn that radical change would be needed. Management was not on board; it did not feel that McGinn was listening. He had strayed too far from those who helped him run the company. O'Shea exhorted his boss to realize that it was essential that top management act as a team and that leadership feel it had a role in righting the ship. He begged McGinn to listen to his lieutenants, saying, "You may or may not be persuaded on any given issue but it's important that they see you are willing to take input from them." The painful truth, as O'Shea explained it to McGinn, was that Lucent's problems were coming from everywhere. He described a ship that had sprung dozens of leaks, in structure, customer relations, new product introductions, supply chain, and order filling. The list goes on. O'Shea told McGinn in essence to go back to square one, to setting expectations as McGinn had done from the very start. In a meeting the following day he exhorted the CEO, "People need to understand your expectations are high AND you are confident they will meet them while at the same time understanding there is no room for failure to line up and get it done." He told his boss not to forget that "this is a team sport."

In August 2000, McGinn reorganized the optical business. McGinn asked Harry Bosco to retire and replaced him with Jeong Kim and Bob Barron. Bosco had been with AT&T/Lucent for thirty-five years; Kim was thirty-nine years old and had come over in the Yurie Systems acquisition. McGinn believed in Kim and saw him as an exceptional talent who, unlike so many others from the acquisitions, was committed to the business. McGinn had witnessed Kim's competitive side at the negotiating table and on the squash court and knew that he was not just a brilliant engineer or investor, but a man with vision who could one day run the company. Kim gave McGinn the report on optical. Development plans were not well managed, the department was in disarray, and the handoff between R&D and manufacturing was the weakest point. The sales force in optical was underperforming, the quantity of business was plummeting, and morale was sinking even faster. The product side of the business had been unable to deliver, and the salespeople felt they had been given an impossible task.

Last year's stars were this year's failures, and the blame was being spread around liberally. Management was not shy about expressing its disappointment with the optical sales force. The sales force, for its part, did not have the hottest-selling piece of equipment in its arsenal. The product development team pointed to Bell Labs' tardiness in coming up with new technologies, while Bell Labs insisted that management had left its development of an OC-192 woefully underfunded. Kim was blunt with his assessment: "As a technologist, this place is heaven, but we'll only dominate optical networking if we execute better and stop being our own worst enemy."

McGinn was desperate to get this business back on track. Later, when reviewing the range of Lucent's troubles, Schacht would argue that optical was a problem that dwarfed others, saying, "Hit optical—you can have this debate about managerial mechanism later." When Kim came to McGinn and asked to borrow his car for the sake of the business, after taking a moment to recover from his laughter McGinn took the request to heart. Financial incentives might help, but salespeople are driven by more than just money, Kim told McGinn; they hungered for competition and public recognition of their triumph. What more public recognition of one's professional achievement could there be than driving the CEO's red Ferrari? Although he never brought it to work, McGinn oh-so-conspicuous car would be hard to miss in the Lucent parking lot filled with American-made sedans and minivans. This was what Kim was counting on.

Kim did not suggest that McGinn could solve Lucent's optical problems with his car, but he knew the move would signal to everyone that when McGinn told his troops that he was "personally committed" to their success, he had done something to prove it. While recognizing that this was only a symbolic gesture, McGinn told Kim that it would show that they were willing to do things differently, to try something a bit crazy, and he began to make arrangements for the car insurance.

At a sales conference meeting with five hundred salespeople and engineers, Kim presented the carrot. As he reviewed the state of the business using a series of overhead slides, he flashed onto the screen a picture of a red Ferrari (not McGinn's actual car but the same model), asking if anyone was interested. The crowd cheered and laughed when Kim put his formal offer on the table. Each quarter, the most successful salesperson would get to use McGinn's car. As long as this person held the position, the car was his or hers. When a new sales leader emerged, the car would change hands. Some were cynical, suggesting that they had little use for a secondhand car, but most eagerly awaited the first quarter's competition.

Although Schacht signaled his nominal support, his management team was not impressed with the notion of loaning out expensive sports cars and offered an array of excuses why such an idea was impracticable. Commissions, they argued, should be enough incentive, and the idea of dispensing flashy automobiles set a bad precedent for other divisions. The whole idea, it seems, had a slightly unprofessional, certainly un-Lucent, feel to it. Kim gave up.

LET DOWN BY HIS LIEUTENANTS

Although McGinn continued to support and promote Aversano, he was extremely concerned about her abilities. Initially Russo, to whom Aversano reported, considered giving her responsibility for all of the Americas. McGinn told Russo that he did not have confidence in Aversano, did not believe she was the most qualified candidate for the job, but that he could present Russo with no alternative. So Aversano was made head of North American sales, a slightly diminished but still vitally important job. With three thousand salespeople reporting to her, she now headed Lucent's largest sales force. Over the years McGinn had promoted and rewarded Aversano, never indicating to her any dissatisfaction with her performance, yet he now promoted her, certain in the knowledge that she was not the most qualified candidate for the job. It was a critical position, and as management often told one another, "As goes NAR, [North American Region], so goes Lucent."

Aversano would now be in charge of some of Lucent's most important client relationships. In giving her such increased responsibilities, McGinn knew she would be working with "very very sophisticated customers, demanding customers, and financially savvy customers, and that was a different set—or an additional set of responsibilities beyond which Nina had previously managed. So I felt we were making a mistake." He was concerned that she would give away margin to get business, that the North American revenue line might increase at the cost of profits. Others on McGinn's team had similar sentiments. "How do I put this? There was an ether, there were words in the ether that people—the executives kind of knew that Nina was struggling. Was I told in an official capacity? No. You just kind of know it in the way of a corporate osmosis," Fitzgerald recalled. Aversano's performance had been discussed in meetings among the top management, and Hopkins had made it clear that in her opinion, Aversano should be fired. She complained to McGinn about Aversano, saying that the job of managing Wall Street's expectations was impossible with a sales manager whose numbers kept moving around. O'Shea and Robert Holder had also commented that the job was too big for Aversano, not just that she had begun to miss her revenue targets but that the increased visibility, management oversight, and leadership were more responsibility than she could handle. Dickson also suggested that McGinn needed to replace Aversano: "In the context of right and wrong people I would take Nina out of her job. She is a tremendous fighter but only sees one way of doing things, i.e. hers. We need a team play[er] to solve these problems as well as an open mind. She provides neither."

Aversano was promoted in the spring. By the summer McGinn's fears had come to pass and he was looking for a way to get rid of her. McGinn remembers that among top management, "collectively—and individually—we had lost confidence in Nina's ability to lead the organization for which she was responsible." And although he discussed replacement candidates with O'Shea, Hopkins, Russo, Holder, and Ben Verwaayen, including Holder himself, in the end they decided to leave Aversano in her position until year-end because to do otherwise would create excessive turmoil at an already stressful and difficult time for the sales organization. McGinn allowed Aversano to stay in the most important sales job at Lucent because he lacked a better alternative. At the time, the decision seemed to have merit. The division's performance was suffering, and a change in leadership might make matters worse. Yet in retrospect (and hindsight is, of course, a beautiful thing), leaving an executive in whom you and your top management have no faith in charge of the lion's share of the company's sales cannot be seen as a wise decision.

When McGinn discussed the matter with Russo, he never mentioned that he was concerned with Russo's performance as well. In order to keep the tumult to a minimum he had kept both executives in place, but the short-term solution had yielded far greater problems. In the spring the board of directors had discussed placing Russo at the head of Avaya, but McGinn had protested. He told his board that Russo was too valuable at Lucent and he could not afford to let her go with the spin-off. Instead he would send his CFO to run the new business and replace him with Hopkins. But only a few months later, in June 2000, McGinn told the board that he was dissatisfied with Russo's performance and that one of the reasons he was reorganizing the company was so that she would no longer have the same level of responsibility. McGinn believed that Russo was incapable of making tough decisions. "She just couldn't force the issue when it was difficult to do things," McGinn recalled. "Because she wanted to work it and create a positive outcome that considered all factors and this [a change he had expected her to make] was something that literally should have been done in weeks, and it took over six months to do it." He found Russo's consensus building to be "difficult when you absolutely have to have a sense of urgency about improving the excellence of the performance of the business." He felt she lacked the sense of urgency that had become more vital to Lucent's business. For McGinn this was an enormous frustration, and, as Schacht recalled, McGinn approached the board in June 2000 to tell them that he had lost confidence in Russo. Yet McGinn never told Russo that her performance did not meet his expectations. Instead, when the two met during the third week in July, he simply described to her changes he was going to make in the organization that effectively eliminated her job. "Then if that's the case, there's really not a job that is comparable to what I'm doing," she said at the time and decided, after almost two decades with AT&T/Lucent, it was time to move on.

Publicly Russo has said only that she left Lucent because her job no longer existed. Yet after her departure a few of her former colleagues, who were still in Lucent's employ, approached her with their concerns. Some felt that the business was being driven too hard and that McGinn's expectations were coming from the marketplace rather than a realistic view of what the company was capable of achieving. Russo remembers that McGinn's leadership was also in question. "I think the market was starting to decline, changes were going to have to be made, and there was just some question about whether he [McGinn], you know, had the team together." She explained that part of the difficulty between her and McGinn was that they did not see eye to eye. "I think it's clear that Rich's management style and my management style were different. Rich did not engage the team as I would have engaged the team. He tended to operate more one-on-one."

When approached by colleagues, Russo did not defend McGinn; instead she said, "I did have a few people talk with me about their concerns about Rich's leadership after I had left, and given the fact that I wasn't in a position—I didn't feel I was in a position to communicate that, I—you know, encouraged the few—a couple of people who raised concerns to speak to either Henry or to the Board." (When Russo gave this explanation in a sworn deposition, her lawyer emphatically refused to allow her to divulge the names of the current Lucent employees she had spoken with, even coming to verbal blows with Aversano's lawyers (in the case *Nina Aversano v. Lucent Technologies, Inc.,* Superior Court of New Jersey, Middlesex County, Docket number L-10004-00), who wanted the names revealed: "I'll instruct the witness not to answer that question." "On what conceivable grounds?" "You can take that up with the court.") Aversano remembers from her conversation with Russo that "I know that Ms. Russo was going to Mr. McGinn. I know after she went to Mr. McGinn what she told me. I also know subsequent to her termination from Lucent that she was extremely distressed and that she told me that she had serious concerns about McGinn's integrity and what he was doing to the business. She told me she felt it was critical that the things he was doing be stopped. And in fact she regretted she personally had not gone to the Board of Directors. She encouraged me to go to the Board of Directors." In her deposition, Russo does not recall making these specific points.

After Russo's departure in August 2000, McGinn met with those executives who had reported directly to her in a conference room at the company's Warren, New Jersey, facility. McGinn looked weary and drawn. Gone was his normal exuberance, the ever-optimistic CEO who had long rallied his troops. "It's not that he was defeated," recalled one executive at the meeting. "His speech seemed almost disjointed under the pressure. With his message he was trying to get everyone fired up, but his tone was different. We had never seen him like this before." McGinn began the discussion by telling the group, "We are not living up to our potential." Russo had left, in his words, because "Pat was not driving change fast enough." Then he looked around the table and said that if this did not change, he too would be out of a job. Each of them would need to make a choice, and there was little time for deliberation. Were they passionate about the business? he asked. He needed their renewed commitment; every person in the room would need to tell him if he or she was on board, and willing to work at a stepped-up pace. He was looking for a personal pledge of their commitment. Some gave him a pledge on the spot; others came to him later privately to renew their allegiance, but something fundamental had changed. "The feeling was that Rich had become overinfatuated with the mystique around Lucent and the mystique around McGinn," recalls one senior manager, echoing what many colleagues said. "I don't know whether I agree with that, but that was the thinking."

In July, Lucent's executives gathered for a North American sales review to develop financial guidance to pass on to analysts for fiscal 2001, which would begin October 1, 2000. Lucent maintains that in Aversano's presentation she assured management that North American revenues would exceed $26 billion, and for the final quarter of 2000 she projected that North American sales would have revenues of between $5.5 and $6.1 billion. McGinn had watched Aversano's numbers bounce around all spring, and he asked her, "Why should I believe you? We have not been able to stay on a number all year." "This is absolutely the number," Aversano told the group. "And there is no reason why we can't have a 25 percent increase next year." Hopkins later explained, "What I recall is Rich being very frustrated that Nina's numbers were—kept changing all the time and couldn't seem to stay in one place." But the problem in Hopkins's view was even larger: "I thought he felt that—and I don't remember exactly the words, that she—it was more than numbers. It was a feeling—I don't want to put words in his mouth. You'd really have to ask him, but that he was uncomfortable with kind of what's going—was going on in the way she was conducting herself."

McGinn took Aversano's numbers and decreased them, based on his recent experience. Yet even the lowered numbers, along with the booming business in Microelectronics, he felt, would get them to their promised revenue goals. This was the reason management publicly stuck to its targets on July 20, when it again promised shareholders 20 percent growth in 2001. At this point Aversano's recollections of the events in July and Lucent's position sharply diverge. Aversano recalls an entirely different situation in July 2000: "I knew that in order to achieve these numbers, we were doing more and more things that were the miracle that were becoming increasingly difficult to deliver. I knew that we had no new software releases coming in in 2001. That any ideas of doing software pooling were gone. We had virtually sold all the software that we could. I knew that we had no financing, all financing deals were shut down. I knew that the emerging service providers market was starting to soften. And I saw signs that the business was going to be more difficult and less profitable and therefore I had growing concerns."

What Aversano failed to say was what she and some of her colleagues, most notably Jay Carter, who was responsible for the AT&T account, had done to inflate Lucent's revenues. In May 2004, following a three-and-a-half-year investigation, the SEC filed securities fraud charges against Lucent and nine current and former Lucent executives. The SEC complaint alleges that Aversano, Carter, and seven others manipulated Lucent's internal accounting controls and "falsified documents, hid side agreements with customers, failed to inform personnel in Lucent's corporate finance and accounting structure of the existence of the extra contractual commitments or, in some instances, took steps to affirmatively mislead them."

Aversano and McGinn have engaged in a bit of finger-pointing over the source of Lucent's ambitious goals. She claims that the overly aggressive targets were set at the top. McGinn disagrees and has said, "Estimates for 2001 were 100 percent bottom-up. There's a dramatic difference between the excellence demanded in a high performance environment and the notion of overreaching." Yet, in truth, neither had much flexibility. The industry was expanding at roughly 15 percent per year; Lucent in its first three years had managed to best this rate and had promised to continue to do so. The revenue figures it needed to continue on this path simply fell out of these constraints. All the finger-pointing in the world will not make up for the fact that the stock market's steep rise and the success of the industry had created expectations that had drastically reduced Lucent's room to maneuver.

Aversano claims that when she told McGinn that she would miss the mark by about a quarter of a billion dollars (or more than a billion from the internal targets Lucent had privately set), he went nuts. Their first open confrontation occurred on an August 14 sales call. The one thing Aversano and Lucent seem to agree upon is that in that meeting, things got nasty. During the call, Aversano told McGinn her division's revenues would not meet the earlier targets. It was only August, and the company was already missing guidance for the fourth quarter of 2000, which it had lowered in July. She remembers that on that call in August, "I told him that at that time, at that point in the quarter we were looking at a view that was 5.2 [billion]. He went ballistic. Basically told me I was going to take down the whole company. And that did I understand I would basically destroy the business if we didn't bring in the 5.7 [billion]."

Evidence suggests that the situation was actually worse than Aversano said in her meeting with McGinn. Only three days earlier she had sent an e-mail to her direct reports: "We have a very serious situation in North America due to upwards of $1B in product shortages. Normally, my reaction to the field is, 'sales people never have everything we need to sell at the time we'd like to sell it . . . we must sell what we have and make up the difference some other innovative way.' However, these shortages have reached crisis proportions, and I am turning to you for your personal engagement and support in addressing this issue. As you know with targeted commitment at $6.3B, at the bare minimum, NAR must deliver $5.7B. Current course unless these shortages are resolved—we are headed for a disastrous $4.2B." A member of her team, Marc Schweig, recounted the August 14 meeting in an e-mail to his boss: "In case you were wondering, Rich was on the Monday morning call and it was very painful for Nina and all of us. He went 'ballistic' over the NAR CWV [North American Region Current Working View]. No surprise. We need to redouble our efforts at any possible deals—no holds barred—that can improve the top line."

When McGinn sat in on Aversano's sales calls, he began to see the reason why things were so unstable. Aversano managed by exhortation rather than analysis. When a salesperson came onto a call and said, "I am not going to do $110 million, I can only do $78 million," the retort from Aversano was not "What happened in the business to miss the mark?" but rather "No, you're not. You're going to do $100 million, or even $120 million." McGinn was losing his patience. Aversano's numbers moved around constantly. At times she was in a panic, telling him they would not get the orders, and at other times she was promising to outdo the revenues he had committed to Wall Street. And she kept using that language, the stream of murky terms that made it clear that more and more her team was looking for "Doug Fluties."

In many ways, these were Lucent's darkest days. Its share price had halved, while stocks such as Nokia's were hitting all-time highs. Again in the autumn Lucent's competitors had all met or exceeded analysts' expectations, and again Lucent looked even worse by contrast. Every other industry player was experiencing unprecedented growth, with rapid rises in their stock prices. While in absolute terms the situation would get far worse for Lucent and the industry, Lucent would never look worse by comparison.

On October 9, McGinn and Aversano met in an office down the hall from McGinn's own office on the third floor of Murray Hill. According to McGinn, when he entered the room Aversano came from behind the desk and sat at a round table with her boss, "and she said that she had made a decision to retire from Lucent, that she's given it a lot of consideration, that she did not want to work in a business where she was not respected by her peers and that she had lots of other things that she was contemplating doing—most likely was going to start teaching—and so wanted to act upon this immediately." McGinn assumed from her comments that she was also unhappy about the relationship between the two of them. Yet he did not immediately accept her resignation but urged her to wait twenty-four hours, giving him a chance to consider other options that might be available to her at Lucent. McGinn recalls that Aversano wanted to know whom she could speak to and he said that such decisions were premature. She told him she had other matters to discuss, and they agreed to meet two days later.

Aversano insists that in their October 9 meeting she gave McGinn a stiff message: she would not privately or publicly support what she believed to be misleading financial information. She claims that she made herself perfectly clear: "Again, I shared with Rich McGinn on the 9th and 11th and again on the 16th that he had to change the perception that existed as to what we were going to be able to do in 2001. That being the market perception. And the analysts' perception. Because North America, which he had always counted on, was not going to deliver 25 percent growth. In fact I told him that it was going to be virtually zero growth." In McGinn's sworn deposition he denies that she voiced such concerns, saying adamantly, "I'm sitting here in this room and I'm looking at Nina Aversano, and I know—and she knows—there was no conversation about that topic [impending sales shortfall] on October the 9th."

Lucent and McGinn deny that Aversano gave them early or adequate warning of the decline in sales, pointing to her as the source of the inaccurate information they later conveyed to analysts. Given the extent of the sales misdeeds the SEC alleges took place, and the fact that such information was hidden by Aversano and others from the CFO, McGinn would be looking at a very distorted picture. As negotiations for her severance package proceeded, Lucent discovered some problems with a deal that had been negotiated by one of her sales staff. As a result, and until they had further information, Richard Rawson, Lucent's corporate counsel, informed her, "You are being placed on paid administrative leave. We will continue to make payments as described in the separation agreement, subject to your complete cooperation in the review." The final agreement between Lucent and Aversano was left unsigned, and when she could not get satisfaction she sued the company under the New Jersey Conscientious Employee Protection [or whistle-blower] Act, and for breach of contract as well.

In her lawsuit, Aversano insists that she made it clear on the July 20 "North American Gap Closure Call" that July revenues were only at 27 percent of their target, despite the fact that the month was more than half over. The numbers were becoming unrealistic because shipments were inadequate and returns were too high. (The SEC alleges that Aversano struck side agreements with product distributors that allowed them to return unsold merchandise, although she refused to put them in writing, according to the SEC. In the case of one distributor, Greybar, nearly everything it acquired at the end of the second and third quarters of 2000 was returned in December of that year.) Lucent, in the words of its lawyers, "denies the allegations" that Aversano let the company know of troubles in the division early in the year. Lucent executives argue that Aversano maintained she would be able to meet her division's targets right down to the wire, in September 2000, when it became clear that this would be impossible. And perhaps even more damaging, to the extent that the guidance relayed to the market was misleading, Lucent's managers have pointed to Aversano as the source of information on which that guidance was based.

Lucent's management felt sorely betrayed by Aversano. She was not some outsider but one of them, a team member from the start. This sense of disloyalty caused real anger among those who run the company, and in the months before the scheduled court date Schacht continued to insist that despite her claims and the mountains of dirty laundry the lawsuit was producing, Lucent would admit no wrongdoing and entertain no settlement. Yet as Aversano's evidence seeped into the public domain, including a damaging speech by Schacht, and as confidential letters and e-mails were exchanged among top executives and depositions taken of Lucent's senior executives, the reputational cost of fighting the suit escalated.

Reviewing the mountains of depositions that were generated for Aversano's lawsuit as well as additional sources, one sees clearly that McGinn felt let down by his lieutenants. Over and over again he is disappointed with researchers and product developers who make claims about products, whether their timeliness or capabilities, that do not come to pass. He feels poorly advised by his scientific staff or sales management at a number of crucial junctures. He does not buy Juniper because of the claims of Bell Labs researchers; he does not forge ahead into OC-192 because the optical division believes the demand is still years away. In both Russo and Aversano, McGinn felt he had executives who were unable to execute and, in Aversano's case, did not convey to him the full extent of the problems in their businesses.

Lucent let Aversano go in a press release on October 16 that described her "extraordinary energy" and "intense focus on the customer" and called her contributions "invaluable." Later, when she sued Lucent, she argued that her case was both classic and tragic, nothing less than killing the messenger. A number of senior executives from Lucent remember things differently. Aversano had been threatening to leave for months, claiming she needed to go because she "could not get the support of her management." They say that in years past Aversano had had a history of insisting that targets were impossible and then meeting or exceeding them, thus making it impossible for McGinn to know where she stood. Far from predicting Lucent's shortfall, the recollections of some at Lucent were that in July and again in August, and in fact right up until September 28, Aversano continued to agree to the $25.5 billion revenue number for 2001. Moreover, they say, she had to know the impact of the deals the company, and particularly her division, was then making upon the following year's sales. "Her saying that she discovered there are lots of credits is like saying that there's gambling going in Casablanca," Lucent's lawyers replied. A Lucent internal document dated September 21 and entitled "Our collective ability to plan and predict performance has been abysmal" shows clearly that Aversano's business, even at this late date, was still slated to bring in $25.5 billion of revenues in 2001.

On October 10, McGinn was back in the hot seat dishing out more disappointment to his investors. Events had moved so quickly and sales deteriorated so rapidly that senior management was fumbling in its ability to discern the source of the problems. McGinn was stunned by the size of the shortfall Aversano was predicting. He remembers, "It seemed incongruous with the enormity of the growth the competitors were describing." He had counted on Microelectronics—which would not be spun off until the following year—to make up for the shortfall in North American sales, but now, with the even larger deficit that Aversano had presented, this would be impossible. McGinn would not meet earnings expectations either for the fourth quarter of 2000 or for the entire year. Although revenues had been expected to come in at the 15 percent level in July (this figure would later be lowered twice), profits for 2000 would actually decline by 10 to 11 percent. This admission was an indictment of Lucent's sales practices. More and more less-profitable business with marginal customers of a lower credit standing was not the strategy that either analysts or investors believed Lucent to be pursuing. Yet here was the evidence.

Reversing the hopelessly optimistic changes it had made in 1999, Lucent increased its reserves for bad debt as it finally fessed up to the escalating credit problem of the CLECs. Despite glaring signs painted across the balance sheet, the company had not wanted to recognize some of the

CLEC payment delays as bad debt. This problem could no longer be ignored. While still insisting that its lending process had been rigorous, Lucent was now acknowledging that many CLECs had enough cash to last only for weeks, rather than months or years, and that for many repayment would be impossible. Lucent would continue to pursue business with what it thought were a substantial number of CLECs that were on a stable financial footing. This was not a total retrenchment, but henceforth the company would be more selective in whom it dealt with.

Two of the most damaging admissions that Lucent was forced to make were that revenues in optical networking systems had actually declined by 5 percent while switching systems revenues were down by 13 percent. Lucent had for four years claimed market leadership and promised spectacular growth in these, its cutting-edge businesses. Although it had always been anticipated that Lucent's switching systems would one day reach obsolescence, the optical market was believed to more than make up for the earnings shortfall. For a year Lucent had been promising to put out the right products and regain the market share lost to Nortel and others. There had never been a boom in telecommunications like the one that was happening in the optical sector, yet Lucent, despite having invented the technology upon which all of these advances were based, could only stand by and watch.

Perhaps McGinn's credibility might have survived had it not been for his continued optimism. Despite the bad news he had delivered for three straight quarters, McGinn continued to give investors and his own board encouragement that the turnaround they had all been looking for was finally under way. He continued to look to his competitors, none of which had slacked off in their sales. This strengthened his resolve. If only he could fix the company's problems, get the right executives into the right roles, get out the hot optical products, all would change. McGinn continued to insist that North American sales could grow at the rate of the overall telecom market. He believed that Lucent was addressing many of the problems that had led to disappointing numbers in 2000 and that it would be back on track in 2001. Even though all of the market's goodwill was gone and it was time to err on the side of caution, Hopkins had promised in July 2000, "We see our way clearly to 20 percent top line and bottom line growth for fiscal 2001." By the October 10 announcement that Lucent would fail to meet these targets there was no longer any discussion of outperforming the market, although Hopkins reiterated her claims made the summer before that she was "sitting on a pile of gold." McGinn and Hopkins promised to give further details on the financial picture, including an update on the explosion of vendor financing and guidance for 2001, shortly after the upcoming board meeting two weeks hence.

Competitors were still insisting that their prospects were undiminished. Nortel's and Alcatel's stock prices reached their all-time highs in mid-2000, as Lucent began to run out of explanations for its failing performance. "We have heard the comments made by Lucent today, specifically those indicating that this is a Lucent-specific issue and not an industry-wide phenomenon," a Nortel spokesman said in mid-October. "We agree with Lucent that the growth prospects remain robust." Alcatel even revised its revenue forecast upward for the year, but this would prove to be wishful thinking. Nortel had already seen its best days, and investors, worried that perhaps Lucent's problems were not confined to Murray Hill, grew concerned about the industry.

On October 16, McGinn again met with his top managers in Warren, New Jersey, to review their plans for fiscal 2001 (although it had begun two weeks earlier) and talk about major changes they would make to the business. The meeting began at 7:00 A.M., and it would be almost 11:00 P.M. before McGinn departed for home. Again at this meeting Aversano's numbers moved. North America would deliver even lower revenues than previously forecast. McGinn remembers that, during Aversano's presentation, there were "audible gasps of disbelief when they saw the material." McGinn was backed into a corner. Many investors had already deemed him to be out of touch with his business, and although he had just warned Wall Street, his predictions were already out of date. In a matter of days he was going to have to tell his board that matters looked even worse than they had six days before.

McGinn had refused to alter his revenue targets because he and everyone else scrutinizing the industry believed that Lucent was operating in a more hospitable environment than was the case. What he could not know was that Lucent was a canary in a mine shaft, an early warning signal in an industry on the edge of collapse. The company's weaknesses made it the first to be devastated by the same forces that would later level the industry. Lucent was the most vulnerable company in its industry. It was selling the wrong products, and its dependency on a few massive accounts meant that when AT&T faltered and Saudi Arabia retrenched, Lucent buckled. McGinn presided over a company that had made great sacrifices for immediate financial results. Lucent had robbed its own future, and now that future was here. Because McGinn could not foretell the future, he would push Lucent to regain its former footing under conditions that now made this impossible.

After Russo left, McGinn began his search for a president. He engaged a search firm to find him a strong operational candidate. For a headhunter it was a particularly difficult assignment. The company was not doing well, the CEO was under fifty-five, and the CFO had huge ambitions to have his job one day. Nonetheless, as headhunters Heidrick & Struggles forged ahead, the biggest question was whether they were looking for a president or for a president and new CEO. On this point, well into October, Lucent's board was emphatic: they were not searching for a new CEO. And as McGinn made his way to the October 2000 board meeting, a pile of résumés for the COO position sat neatly stacked on his desk.

The pressure on Lucent's board to take action rose daily. McGinn had been unable to keep his promises, and many wondered if he should not be shown the door. Press reports and analysts' comments all focused on the timing rather than the possibility. The board was widely criticized for being out of touch with the industry ("fiddling while Lucent burned"), overpaid, too small (only six members, rather than the average eleven), too old, and too slow. They had even managed to earn themselves a place on *Fortune*'s list of the "Dirty Half-Dozen," or the six worst corporate boards in America.

There are no clear-cut criteria for firing a CEO, no set of rules that can be easily relied upon. McGinn served at the pleasure of his board, and at no time had he held a discussion with them about the possibility of his leaving. As he headed to the October board meeting armed with more bad news, a number of his colleagues in senior management speculated among themselves that his career with Lucent was over. Yet McGinn knew he held the board's confidence. They might give him a wrist slapping or even put in place some tighter controls, but private conversations had convinced him that his position was secure. Schacht had stepped off the Lucent board in early 2000, but when McGinn could find no one to replace him, he had invited Schacht back. Lucent was not doing well, and McGinn's offers of directorships were not getting even a nibble. One senior executive says, "I remember on the day that Schacht went back on the board thinking, this is a problem for Rich."

Privately one board member had told McGinn over the weekend that although the board was troubled, they would get through this and McGinn would not be asked to leave. Only days earlier, on a conference call with four hundred Lucent executives, McGinn had insisted that he had no plans to resign. He had no conversations with board members before the meeting about his possible departure or perhaps being allowed to stay but on a shorter leash, with less ability to maneuver and more direct control by the board. Yet McGinn was not naive about his plight. Among his inner circle he had been more open about his despair. Richard Rawson, Lucent's general counsel, remembers him saying that a "board tires of a CEO that doesn't make his or her numbers."

At one Saturday morning session, October 21, the top executives were meeting with the board when board member John Young asked where the reports on each of the business units were, as he could not find them in his notebooks. There was a copy in McGinn's notebooks, but the papers had been inadvertently left out of the other board members' binders. Franklin Thomas, the former head of the Ford Foundation and a board member whose judgment and advice McGinn had often sought, suggested that rather than run out and find a copy machine the various business leaders might brief the board on their plans over an informal lunch. McGinn concurred and asked each of his executives to make a presentation. His business leaders were caught by surprise. As they were given a chance to speak directly to the board, the results were damaging for their boss. None of them had prepared a presentation, and as the meal progressed, the board probed deeply into the problems plaguing their businesses. During this lunch a number of business heads laid out for the assembled board the extent of the troubles that Lucent was facing. They told the board that employees no longer had confidence in Lucent's leadership.

During the weekend board meeting, Hopkins painted an even more pessimistic forecast of Lucent's financial outlook than she had delivered only two weeks earlier. Wall Street was still looking for earnings growth, but there would be none. There was no gold mine, and in the current quarter she was expecting a decline in revenues. The board was shocked to hear this most recent round of bad news. In an emotional confrontation, the audit committee raked Hopkins over the coals. How had the numbers fallen so far, so fast? Why had she and McGinn not known about them? Saturday evening, the board, minus McGinn, met in closed-door session for four hours to discuss his fate.

Perhaps McGinn had underestimated the degree to which he had lost the board's confidence. He was not particularly cozy with his board, as Schacht had appointed all but one member and relations with some members had deteriorated with each bout of disappointing news. His relationship with Paul O'Neill had never been very good and had only recently worsened. O'Neill was angered at the newest round of weak numbers and berated the senior team for its performance. McGinn was very unhappy with the confrontation and after his executives had left the room told O'Neill that he was having a hard enough time holding on to people under these conditions. If O'Neill wanted to attack someone, it should be McGinn, because telling the senior executives that they were not worthy of the Lucent name was not helpful.

Behind McGinn's back a number of senior managers had been discussing the weakness of his performance privately with board members. One senior manager said, "I remember being on a boat trying to get a cell phone connection. I called Betsy Atkins [a board member] and said things were out of control and McGinn had to go." One of Lucent's board members remembers this more as a coup led by Schacht than a sorrowful recognition that a CEO had underperformed. Schacht, according to one member's recollection, told the assembled board that in his discussions with managers at the company he had come away believing that McGinn was disconnected from the problems of the company and not sufficiently communicating its difficulties to the board. The two stories are not inconsistent. Schacht was the board member with the closest knowledge of the operations of the company, and managers would very likely have gone to him with their concerns. Schacht later explained that he went into the weekend prepared to continue to support McGinn, that he had believed McGinn would remain as CEO, but as the new revelations had seeped out over the weekend, his stance had changed.

Sunday morning the compensation committee met. The purpose of the meeting was to hand out additional options to the senior staff in order to try to hold on to them. Recently Lucent had had the experience of offering hundreds of thousands of dollars in options to employees who had come to the company through acquisitions, only to find the offers declined. Again O'Neill and McGinn squared off. O'Neill thought the senior executives were being greedy. The business was foundering, and yet the committee was considering rewarding those who had failed to perform. McGinn told him he needed the options to hold on to people who were inundated with other offers. After this confrontation the board met alone without McGinn.

Later Sunday morning, the board gave McGinn the news. Schacht and Thomas were the bearers of bad tidings. The board had lost confidence in him and was looking for a CEO with a different set of skills. McGinn had shown himself to be a strong strategist, but they needed someone who could execute those plans.

For McGinn it was the end of thirty years of loyal service. He felt that he had been ousted for transient factors, including a readily acknowledged product mistake and cutbacks by two large customers that were beyond his control. While at each step along the way he had taken responsibility for the company's troubles, he also felt he knew what was needed to fix the problems. To this day McGinn does not believe that he was overreaching Lucent's abilities in the goals he set in 2000. From the board's perspective, it had a chief executive who continued to promise and continued to disappoint and did not have the confidence of his organization. It had a leader it believed was out of touch with the realities of the marketplace, had established and clung to goals the company could not meet, and had set a tone at the top that had allowed sales practices to deteriorate. For Schacht, it was the end of a relationship he had described as unique. "He's a friend of mine and I admire him immensely," he said shortly after McGinn's dismissal. "I can't imagine the pain he is going through."

After a brief conversation with Thomas, McGinn simply walked away, never to return to his office or to any other Lucent facility. He met briefly with his senior team, some of whom were tearful, and told them that he had all the confidence in the world that they would do great things. Controller Jim Lusk remembers speaking to him shortly after the board's dismissal: "McGinn took full accountability. He said, 'The board has made their decision. I am the chairman, so I am accountable. I don't want you all to lose faith in the company.' And he told us he would be all right and never tried to blame the board or anyone else." Later McGinn sent all his former employees a heartfelt e-mail telling them that he had "been the luckiest man in the world." He said that his life had been enriched personally and professionally for having known them, and that when industry spending turned around they would once again accomplish great things. He told them, "I have gained so much by being with you that I am focused on that and cheering you on from a distance."

CHAPTER NINE 2001

2001 Stock price   High: $17.26   Low: $4.09

> *Our foundations are sound as our restructuring program starts to turn us around: There is no crisis pending.*
>
>   Kathleen Fitzgerald
>   longtime Lucent spokeswoman

> *In spite of considerable effort and success in several areas we are a company in crisis.*
>
>   William O'Shea
>   internal memo to Rich McGinn

On Monday, October 23, 2000, Schacht stepped seamlessly back into his role as CEO of Lucent, albeit on a temporary basis. At his second inaugural address to employees his sense of humor did not fail him, and he played a clip from *The Godfather III,* a scene where Al Pacino says, "Just when I thought I was out, they pull me back in." The change of leadership was met with universal acclaim, and for Schacht there was no transition period, as he simply returned to his old office. Schacht personally knew the CEO of every major customer and within hours of his reappointment was on the telephone providing explanations and reassurance.

It was a homecoming and a new beginning. Schacht was given a hero's welcome when he returned to the employee cafeteria. Wall Street rejoiced at McGinn's departure, despite the fact that his replacement was the man who had mentored him, sat on his board, and backed up his decisions for more than two years. From the analysts' perspective, Schacht restored Lucent's battered credibility; during his tenure, the company delivered on its promises. Schacht knew things would need to work differently this time around. "We developed a reputation for calling out quarters that we were then not able to meet," he told those assembled at Lucent's first annual meeting after his return. "We're trying to talk less and perform better."

A change like this does not come without a barrage of questions, and they were all aimed at Schacht. What went wrong? Who was to blame? Why was McGinn given so many chances? Schacht took his share of the blame, pointing out that as the former CEO and a board member throughout Lucent's lifetime, he could not avoid it. He felt that the board had acted correctly but admitted, "In retrospect, when you change your CEO, you always wish you'd done it sooner. But does a guy who's delivered that kind of performance get to miss a quarter? Darn right. Would any board, having been delivered that kind of performance, move sooner? I don't know."

Responsibility lay with Lucent, according to Schacht. The pressure for growth, as he saw it, was "self-induced." In hindsight Schacht felt that McGinn had lost sight of the long-term interests of the business, saying, "I think my judgment was that Lucent became intoxicated with Wall Street's expectations." The new CEO agreed that market conditions had been anything but typical, but felt that Lucent (and by direct implication, although without stating it, McGinn) should have cast a more jaundiced eye on the demands from Wall Street and its messengers, the analysts. "Stock price is a byproduct, stock price isn't a driver. And every time I've seen any of us lose sight of that, it has always been a painful experience," he said.

In his push for growth, Schacht argued, McGinn had overestimated the company's ability to expand. "I think we tried to run faster than we were capable of. It was a valiant try. I applaud the attempt. It didn't work," he commented on his second day back. Lucent had simply not been able to operate at breakneck speed. The infrastructure of the company (working capital and cash flow management, inventory and control systems, and the like) had been unable to efficiently process the additional sales, meaning that costs had ballooned as management scrambled to put systems in place to handle the increased volume. To his grave disappointment, Schacht found out in his first weeks back that the most basic processes of the company had simply broken down.

After a brief analysis Schacht determined that Lucent's problems were "not on development or product or on the side of understanding the market, understanding the strategy or understanding where we want to be," adding, "What we need to do is accelerate and bring focus and deal with some of the systemic issues that are before us. I think you can measure this by our performance over the coming months. For reasons that are not yet fully clear, we have fallen behind in the execution arena. But I would rather fix execution issues than strategy issues or product issues or people issues."

In this analysis Schacht may have initially underestimated the extent of Lucent's difficulties. Lucent did have people issues. Many of the staff who had arrived from start-ups had vested their shares in 2000 and were now fleeing. Those more loyal to Lucent were deeply demoralized, anticipating that job cuts were in the offing. "The parking lot fills at 9 o'clock and empties out at 4:30. Engineers play chess all day," said one former optical networking manager who later departed. "There's no sense of urgency, no sense of accountability. They are all looking for new jobs."

And Lucent certainly had product issues. Its own chief strategist laid the blame for its failures on product problems. "It all boils down to kind of one fundamental thing," William O'Shea explained. "And that is, we made a decision around OC-192 and 10-gigabit optical systems, that frankly, we're still living with. At that point in time, that stuff was just getting started, and we and Nortel were just about neck-and-neck in optical. Nortel is now several times our size. If we had made a different decision and had shown an optical business this year and this quarter the size of Nortel's, we wouldn't even notice the other things we talked about."

Lucent's strategy would also soon be shown to have failed it. Lucent had gone for Plan B. Management had assumed that the CLECs would prevail, then overhauled the company to meet the newcomers' needs. As its financing, sales effort, and product mix had shifted toward CLECs, Lucent had become more reliant on less stable customers. But by 1999, CLECs had garnered only a tiny portion of the business data and telecommunications market and only one CLEC was profitable. The Baby Bells after a period of retrenchment had proven to be aggressive competitors willing to invest in new networks, cut prices, and take on the competition to defend their markets. Without fresh funds from Wall Street and venture capitalists, the CLECs would be gasping for cash. By 2002, only seventy of the three hundred CLECs that had been in operation in 1999 were still in business. And their assets, far from being an inflated multiple of the amount invested as most had once expected, now fetched only pennies on the dollar. With the clarity that only hindsight affords, the strategy of aggressively lending to these now-bankrupt businesses could be seen as a horrendous mistake. Lucent, and the industry, had focused its business in the wrong place. As Lucent altered its strategy under Schacht, the entire orientation of the company needed to be quickly redirected and redeployed toward the most stable sector of the industry, their old friends the Baby Bells. These factors explain why Lucent began its decline earlier and more dramatically than any other company in the industry. But ultimately it was the industry's collapse that undid Lucent. If telecommunications had continued to grow, after some period of painful dislocation Lucent probably would have recovered.

On one of his first days back at the job, Schacht met with top Bell Labs people. His spirits upbeat, he conveyed both enthusiasm and control. He told the scientific team, "I had to save my own company [Cummins Engine] twice, and this turnaround looks even easier than Cummins." Saving Lucent, he told them, was a matter of hard work and getting back to basics. What he failed to tell them was that restructuring Cummins had taken the better part of a decade, he had sold 20 percent of the company, and that for four years the company had lost money. Nonetheless, these comments inspired confidence among those who heard them. But neither Schacht nor anyone else had the slightest inkling of the tsunami that was about to break over the entire industry.

As he had done when Lucent was just starting out, in early November, Schacht took some fifty to sixty of his senior officers off site for two days to the Parsippany Hilton. Public Relations vice president Fitzgerald was pressed for time in arranging the meetings and was able to book only a cramped slice of a large ballroom, a long, narrow space with poor acoustics. Despite the sensitivity of the information he was about to reveal, Schacht asked Fitzgerald to tape the proceedings. To address the group, he stood at a small podium in the middle of the gathering. There would be two days of hard, sometimes painful work. Schacht explained that the first day they would look backward, focusing on the lessons that might be learned. After that, "We're going to talk about values, and we're going to remind ourselves about our value statement. And we're going to make a reidentification and recommitment to the value statement that is the driver of our everyday behavior."

Schacht set before his team the harshest truths. For months Lucent managers had come to him as the former CEO and board member to tell him of the company's travails, yet it was only in the past two weeks that the full scope of the problems had been fully revealed. The truth was ugly. "Just so we all understand," Schacht warned. "Ex-micro [without Microelectronics, which was intended to be spun off shortly but was delayed until 2002] we lose money in the first quarter, our largest quarter. So, when we talk about crisis, and we talk about turnaround, these are not overstatements."

Schacht began where he had started five years earlier, talking about the importance of "we," and scratched a note at the bottom of his speech to remind himself to reemphasize to his audience that "there isn't any 'they.' " He touched upon the values statements that they had so carefully forged together and reminded them that an organization tethered to such solid beliefs would not go adrift. While Schacht hoped to be forward-looking, he was going to take an excruciating and honest look at the past. Schacht had debated with himself about publicly owning up to the failings of the recent past. He felt it was a risky approach, one that did not "make very good reading." But the risk of not going over this painful territory might be even greater. Everyone needed to understand the events that had occurred and walk away with the same lessons. It might not look pretty, but he wanted to make certain everyone agreed upon what would be different going forward.

All that had transpired under McGinn was labeled the "old model," and Schacht drew a line in the sand. This would be the day to shine a glaring light on all that had gone before, and then it would be over. Schacht asked his employers to hark back to an earlier time when they had been in top form, meeting their "best-in-class" objectives and getting back to the type of performance that he had presided over. He sent them off in groups of ten to consider the lessons learned. To some his words were a huge comfort, signaling that the craziness of the past eighteen months was over. Schacht's moral compass and enduring values had successfully led the company and would again. To others his speech was devastating. He was repetitive and offered palliatives rather than specific actions with time lines for implementation attached. He insisted they were operating in a strong and growing market, but to many executives in daily communications with their sales teams in the field, this did not ring true. Going backward could not be the answer, and as one executive pointed out, "He might just as well have worn a smile button with his old phrase, 'Opportunity of a lifetime,' written under it."

While implicitly Schacht was laying the blame for Lucent's troubles at McGinn's feet, he never mentioned his former protégé's name. And very much in keeping with his philosophy of "we" and his self-effacing management style, he told those gathered, "I'm part of this too. I'm part of the old model, I don't like it. This isn't scapegoating, it's not you and I'm up here lecturing. I've been part of this too. Okay?"

Schacht did not pull any punches. The short-term decision making that had enveloped the company had been at the expense of long-term needs. "What has happened to us is that our execution and processes have broken down under the white-hot heat of driving for quarterly revenue growth," he said. In two weeks of discussions with colleagues he had come to see that the all-consuming focus on quarterly revenues had led to some disturbing behavior patterns that resulted in practices that sometimes "moved uncomfortably close to the edge of respectable behavior and in many cases, or at least some you've shared with me, well-beyond the mid-point of our value set." The SEC would later charge that Lucent and nine of its employees "with knowledge or recklessly" engaged in fraudulent practices and told lies in accounting for the business the company conducted. Schacht is not a leader who skirts the details, so to emphasize his point he rattled off a list of practices all of which had helped to bankrupt the company's future. He publicly exposed a list of sins: "Gap filling became a way of life. Pull-ups from month one and two to fill upcoming month three. Selling forward at discount for current quarter. Buying business at large discounts. Channel creation. Bill and hold. Expensive vendor financing. Stretching revenue recondition practices . . . Putting off write-offs and using the balance sheet for revenue generation . . . And a controls and information system which was strained at half our current size." Schacht readily admitted that he had not known the full extent of these practices until he returned to Lucent and that over the course of fiscal year 2000, nothing had been done to make shareholders aware of the disintegration of business practices that had taken place there.

Lucent's customers were unsettled by the turn of events, and what Schacht heard from these customers in his first weeks back was a damning indictment of a company that had sworn on the bible of customer service. Major customers, Schacht relayed, had said to him, "You can't keep your promises. Your shipments are not on time. They don't work up to the standards that we expect from you. You're too hard to deal with. You're late." Another customer with whom he had a long-term relationship simply begged, "If it would help I'll get down on my hands and knees, please get your act together. You can't imagine how bad the Cisco equipment is, and it pains me to tell you that it is better than yours." Lucent's problems had become its customers' problems, which in turn had become the end users' problems. Faulty software caused network outages that enraged customers. The most painful revelation had come from a client who told Schacht, "Compared to competitors, Lucent is consistently least creative in its proposed solutions. Most of the highest priced solutions, the most bureaucratic and rigid in negotiations. Has not been willing to offer logical solutions to our products." Then the customer threw his staff out of his office and sat down alone with Schacht and privately bared his concerns: "I don't understand what your guys have been doing. I've gotten the craziest phone calls a month before the end of the quarter for the last three quarters that I don't understand. I got so concerned I asked our law department to come in." The legal opinion said, "Look, that [giving the customer a good deed] increases our shareholder value. If they're [Lucent] crazy enough to do it, that's their problem, not ours." The customer continued, "Henry, that is no way to do business with us."

One of the most dramatic admissions came from Jay Carter in a story he told in front of the entire assembled group. Carter was a corporate officer and president of the AT&T account, and as such was one of Lucent's most important sales managers. One evening in early January 2000, Carter was at home watching a football game. A senior manager at Lucent telephoned to ask about the status of a software buyout, a deal that Carter had in the works. In the course of their conversation, Carter gave him the bad news. Although the terms of the agreement had been settled, he had been unable to book the deal on or before December 31 and the revenue could not be tallied into the first fiscal quarter. The senior officer told him to keep working on the deal and gave Carter the clear impression that if he was successful in completing it anytime soon, the sales would find a way into the previous quarter's tally. Carter told the assembled executives that he was pressured to do what he clearly knew to be illegal, and although he had resisted (the deal was booked in the quarter in which it occurred), it was evidence of a change in tone at the top. As one executive remembers, "It was an environment where you were asked to lean but not step over the line." Hopkins stood up and complimented Carter on his candor and courage in telling the group the story. A postscript to this anecdote is that the SEC has accused Carter of taking steps to "mislead Lucent's Chief Accountant about the existence and nature of the side agreement with AWS [ATT Wireless Services]." The SEC has asked the Court to force Carter, Aversano, and the seven others to pay civil penalties and disgorge their "ill-gotten gains."

On the second day, Schacht told his executives a story. Over the course of the October board meeting, as it became clear that McGinn would leave, he wondered if he could return to Lucent. He spoke to his wife, telling her that he had already retired twice, once from Cummins and once from Lucent, and did not know if he could go back. She told him emphatically, "You had the opportunity to build a great company and now you need to go back and fix it." His audience applauded this tale. One senior Lucent executive explained, "He [Schacht] thought McGinn messed up his business and now he was going to come back and save it."

Some who were at the two-day meeting remember it as the closest thing to a tent revival that they had ever witnessed. After Schacht's long-winded twenty-two-page speech, he broke the executives up into smaller groups with some instructions: "So first session, what are the lessons learned? This is not rocket science. Somebody take notes, we'll go around and we'll get them on the board. This is blurt it out time."

In each group executives recounted their transgressions, relaying the details of deals that had been too aggressive, not in the long-term interest of the business, and confessing to their sins. One by one they stood up and told of transactions that they should have walked away from, times they had not had the backbone to refuse, when they had not acted as steward for the business. They told how in their heart of hearts they had known these were bad practices and resolved to behave differently going forward. Aversano, whose retirement had already been announced and under whose management many of these infractions had occurred, sat in the midst of those reviewing their past. While McGinn's presence hung over the gathering, he was not mentioned. In this mass mea culpa, the only finger-pointing was in the form of "management made us do this," but other than McGinn and Russo, management was all sitting right there in the Parsippany Hilton.

Schacht turned his focus to the customers and the details of running the business. He believed that most of Lucent's problems lay in the details (the industry's undoing still lay in the future), and that if he could fix these operational problems the company's fortunes would improve. In the months after Schacht's return, a senior colleague recalls, "Henry's earnestness was a welcome respite from Rich's brashness." Once he had stabilized the situation and restored confidence at home, Schacht took to the road. One senior sales manager remembers this as a break with the past. As the company had become mired in its own troubles, McGinn had sometimes been difficult for his sales leaders to reach. Schacht, on the other hand, became immediately involved with many of the major sales efforts. When trying to close a deal with Global Crossing before the Thanksgiving weekend, Schacht make it clear he wanted to be kept abreast of the developments, that he was available to make phone calls from home over the weekend in aid of winning the business.

In one meeting with Winstar's CEO, William Rouhana, the CLEC sought additional funding. Schacht was not interested; he was not at this meeting to do deals, he explained, but rather to discuss Lucent's new direction. Winstar's management was not pleased and told him, "We've done you favors." To Schacht, this was an extremely worrying phrase, and, when he returned to the office, he ordered a review of Winstar's account to begin that day. He told colleagues that he had not liked the sound of Rouhana's message and he wanted Lucent's lawyers and accountants to take a hard look at every transaction with Winstar. He asked if there were any other accounts that might look like this one. If there were any problems, he wanted them uncovered. Over that weekend the outsiders found only one deal that was not in their view legitimate.

Delving further into Lucent's sales practices yielded more deals that were not as they should be, and, while not admitting any illegality, Lucent voluntarily restated its earnings for the quarter and reported the questionable issues to the Securities and Exchange Commission. The SEC is clear on what it considers adequate criteria for revenue recognition. Its rules list four criteria: "evidence that an arrangement between the buyer and seller exists, delivery of a product or rendering of a service, a set of determinable prices and an assurance that payment can be collected." Specifically, Lucent reduced revenue by $125 million because a sales team had improperly extended a credit to Winstar. The company pared back revenue by a further $74 million to reflect an arrangement with BellSouth that had been improperly recognized. Lucent also decided to take back $452 million for equipment that distributors had failed to resell or install and another $28 million for a system that had been incompletely shipped. In 2004 the SEC accused Lucent of improperly recognizing more than $1.1 billion in earnings, thereby falsely increasing 2000 income by 16 percent.

In the final days of the fourth quarter, problems had developed in another software license deal, this time with BellSouth. After months of hard work, the deal looked unlikely to close by year-end and alternatives were considered. Lucent had offered BellSouth a software pool whereby the carrier committed to purchasing software in the future from a "pool" of different products (and in some cases paying only a token amount up front) and booked the revenues into fiscal year 2000. It was simply a mechanism to move revenues from 2001 into the earlier year. The Lucent executive involved with this deal expressed her misgivings to

McGinn and Hopkins when she wrote, "Giving them the incentives for a $75 million software pool which merely pulls forward software revenue from future quarters is neither good business nor a good precedent to set."

Schacht was understandably defensive about any accusations of illegality. The deals that needed to be restated fell into that gray area of legal and defensible but inadvisable, and he used this chance as a new CEO, even a returning one, to wipe the slate clean. Schacht promised the market whiter than white, saying, "We're going to restate everything that's even close." He was inundated by a barrage of questions about criminal wrongdoing in Lucent's sales force and immediately defended its actions and those of his management: "In fact, it's important to remember it was we who found this and it was we who decided to take the restatement of revenue. We did our own investigation. We looked at every single revenue-recognition item. We did this with an independent outside counsel and with another group of auditors called forensic auditors—what a marvelous phrase—and they, independent of us, went through all of our books. No elements of wrongdoing were found. There were some judgments that we believe, in retrospect, should not have been made. And the people who made them are no longer with us."

Still, by restating a total of $679 million in revenues (which would be only a portion of what the SEC later alleged), Lucent publicly recognized that its sales practices had not been all that they should have been. One senior regulator commented to *Fortune,* "Wow, $700 million. I think you could see some people ending up in striped suits in this one." The problem with admitting to misdeeds in a single quarter was that the observer could not help but wonder what had happened in other quarters. Hopkins was keenly aware of this risk and in a late-night e-mail to Schacht shortly after Lucent admitted the accounting problems told him, "I decided I needed to read the coverage. My concern is that this is coming off as though we can't do accounting vs. the situation where we [were] not given the facts. Both the *Journal* and *Times* hit me personally with pretty negative comments. I'm getting lots of calls from friends that think I've lost my mind doing aggressive accounting . . . and that's friendly fire. Any thoughts on how we can direct this at all? I'm all for character building but this is getting a bit exciting."

In order to achieve its lofty targets, Lucent had slipped step by step into increasingly lax revenue-recognition practices. Standards had begun to erode in 1999, and by 2000, sales teams had become extremely aggressive in creating deals with their customers. Looking back at events, Schacht reflected, "In striving to sustain growth targets, there's no question that we made some deals with customers that had terms we are paying for now."

THE COLLAPSE

Back in late December, Schacht had begun by ridding everyone of any lingering notion that Lucent would make money in the first quarter of 2001, just days before the quarter concluded. It was the third time Lucent had given its investors a revised outlook for the quarter, a symptom of the warp speed at which conditions were deteriorating. The previous summer, shareholders had been told that Lucent's revenues would grow by 20 percent in the first quarter of 2001. Now, only months later, Schacht was telling them to look for a loss of equal magnitude. While the financial results of 2000 seemed devastating, as net income had fallen from $4.7 billion in 1999 to $1.2 billion in 2000, no one could have guessed that in the year just beginning Lucent would lose more than $16 billion. It was a swing of epic proportions.

For the first time, Lucent's management willingly admitted the source of its problems. Sure, OC-192 had been an ugly mess and the year would have looked better if Saudi Arabia and AT&T had stuck with their original spending plans, but the biggest problem, as yet unarticulated, was that the market for telecommunications equipment was collapsing at a speed that stunned everyone. Schacht pointed specifically to the emerging problems in the CLEC market, a slowdown in spending by Baby Bells, and a drop in software sales. It was a difficult message to deliver, for two reasons. First, it contradicted the information the company had been giving all year long, that the difficulties resided at Lucent alone. Second, competitors like Nortel and Alcatel, after watching their stocks hit all-time highs over the summer of 2000 and remain elevated all fall, were still predicting growth rates of above 25 percent for 2001.

In the autumn, Nortel and others continued to insist that growth in

2001 would be robust. With its booming optical business, Nortel suggested that its revenue growth might even reach 40 percent during the year. Yet by February 2001, both the economy and the industry showed real weakness, and for the first time in thirty-two quarters Cisco missed its numbers. Nortel was forced to issue two profit warnings in the second quarter, explaining that the collapse had now spread globally. But when Nortel and Cisco lowered their guidance in February, Lucent did not follow suit. With optimism that would not be rewarded, Schacht ventured, "It may very well be that we took our hit in the first quarter. Right now our forecast is holding firm for better sales this quarter than last quarter."

Despite Schacht's assurances about a return to profitability, Lucent's destiny had slipped beyond his control. It would be a year of restructuring, Schacht asserted, a year when Lucent's foundations were repaired and strengthened. As every CEO attempting a turnaround likes to say, Lucent would be leaner and meaner. The company took a onetime restructuring charge of between $1.2 and $1.6 billion (this would be tripled by March) for reducing head count and writing off the cost of some of its acquisitions. Expenses were reduced by $2 billion and head count by ten thousand, while even more manufacturing was outsourced. Working capital the funds tied up in the day-to-day workings of the company came down by $2 billion, capital spending by $400 million, and new credit facilities were established, secured by the company's assets. Most important, the entire company would return to its AT&T roots by focusing on the thirty largest customers. The goal was "to do it once and do it right." Schacht promised "sequential improvement" in both revenues and profits over the course of 2001. The plan was eminently sensible, and in any year but 2001 it might have worked.

Lucent did not emerge from 2001 strengthened. Investors who stuck with Schacht, assuming that his past pointed to a profitable future, were rewarded with a loss of 95 percent of the value of their shares, not from their all-time highs, but from when he took over. In November 2000, Lucent was at $24, but by July 2002 it sold for less than one dollar. For employees, most of them longtime loyalists, Lucent's head count continued to be cut as the company struggled to regain financial footing. In an excruciatingly painful process, Lucent's management slashed the

Relative Performance, Lucent v. S&P Index, April 1996–January 2004
(04 April, 1996 = 100)

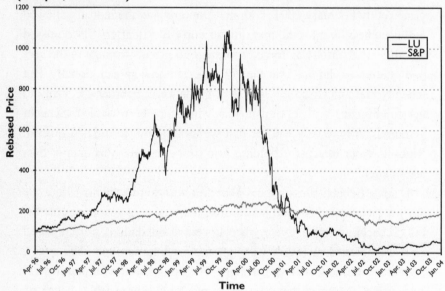

Data courtesy of Iverson Financial Systems.

number of staff (through spin-offs as well as layoffs, both voluntary and involuntary) in an effort to seek financial stability. In 2000 Lucent had 150,000 employees worldwide. By the end of 2001 this number had been reduced by more than half, to 62,000, and by late 2003 it would be cut by nearly 75 percent, to 36,500.

The collapse of the telecommunications industry, as capital expenditure plummeted and demand for Lucent's products evaporated, meant that Schacht's plan would not be enough. The only way for Lucent to regroup and recover would be to sell off assets, using the infusion of cash to reduce the company's indebtedness. From the first days of his second term as CEO, Schacht could see that Lucent was going to run out of cash. Bank loans expired in February, and Lucent would have a difficult time when it sought refinancing. Avaya was gone, leaving Lucent with four key assets to sell: Agere (the former Microelectronics), Lucent power systems, Optical Fiber, and any pieces of the acquisitions that held residual value. Many of the businesses that were not sold were shuttered, even if they had promising products, in an attempt to control expenses. Times would get so tough that Schacht would sell Lucent's jets and cut back on gardening, bottled water, late-night pizza, and light-bulbs—not every lightbulb, but every second or third.

Firing staff was inevitable, but it was something Schacht was loath to do. Throughout the downturn, layoffs were made only as a last resort. Schacht, like most of his employees, came from a world where the company that hired you out of school was the one that paid your pension. Even as Lucent's fortunes became more difficult to predict and incoming cash dwindled, Schacht kept job losses to a minimum. From a human dimension this was deeply admirable, but it meant that Lucent never got its costs below its break-even point, and over time the waves of job losses continued. The upshot was that Lucent remained too highly staffed for the amount the company was earning even as the business ground to a crawl in 2002. A number of Schacht's top executives have suggested that he should have been more aggressive in reducing Lucent's head count more quickly. Their argument was that if Lucent could reduce its operation and conserve its cash, even if it meant cutting a bit too far, the company would be in a position of strength when the telecommunications market turned. Instead, Lucent stayed behind the numbers, each quarter announcing unanticipated losses in jobs and revenues as it continued to burn its cash, credit rating, and credibility.

Schacht proved no more able to predict the company's future sales than McGinn had been. After announcing that Lucent had identified those revenue-recognition issues affecting revenue for the fourth calendar quarter of 2000, the company withdrew guidance for the foreseeable future. Both the industry's conditions and Lucent's situation were changing so rapidly that, although the quarter was two-thirds over, management could not guess what results would look like. Schacht continued to "call out quarters" (as he had criticized McGinn for doing) and failed to meet them every time. He promised a return to profitability in 2001 and then in 2002. Each time he set a target date, a specific quarter, and each time he missed the mark. Each time Lucent's management followed through with its plans, making the announced cost reductions, and each time it was not enough. Within months, sometimes within weeks, every promise was retracted with further explanations about how the industry was in deeper trouble than previously thought. This was to be not a spending slowdown but a total retrenchment, the likes of which the telecom industry had never seen. In 2001 telecom spending fell by 10 percent, but by 2002 the decline was an astronomical 43 percent. While it was entirely true that the crash in telecommunications shattered even the most pessimistic forecasts, at no point did Lucent come close to getting ahead of the decline rather than being a victim of it.

In many ways, Lucent went backward to go forward. Executives from the acquisitions who had filled top spots under McGinn left, and those who had come to Lucent together from AT&T filled all the most senior positions. Schacht believed that the task of overhauling the company was so great that it could be undertaken only by those who knew it best. The looming cash crisis caused by credit that was due to expire on February 22 accelerated Lucent's timeline and, for Schacht, argued against bringing in outsiders. He believed that he needed sign-off by his whole team on a plan for consolidating the company, and that people who already knew one another well could best accomplish this. He simply did not have the time to bring outsiders up to speed as his bankers breathed down his neck. Every quarter, Lucent shed tens of thousands of staff as it announced disappointing earnings. A number of Lucent's most advanced innovations, like the Lambda router, would be shelved for lack of demand. The company returned to its roots, refocusing on entrenched players, the Baby Bells and AT&T, which had not been the central focus in Lucent's drive to meet CLEC demand. Lucent had little choice but to pursue this strategy, as it looked as though its traditional customers would be the last ones standing. In the end, three-quarters of Lucent's staff would be gone, revenues would shrink to a quarter of what they had been in 1999, and with many of the businesses shuttered, spun off, or sold, but for the changes in technology Lucent would look much as Network Systems had in 1990.

"IT'S BAD AND THERE IS NO CAVALRY COMING." —SCOTT CLELAND, TELECOM ANALYST

The debate about whether the United States was in a bear market had been a lively topic in 2000, but by 2001 it had turned into a discussion of how long it would last and how bad it would be. The laws of physics gave cause for concern; the notion of an equal and opposite reaction, in both velocity and magnitude, was a troubling prospect. Still, there was a prevailing belief that, like other industries, telecommunications would feel some pain, but as an industry fundamental to the economy, not the speculative froth that was the dot-coms, it would stand up better. It took most of 2001 to wipe away any lingering notions that telecommunications had not been the biggest speculative market of them all.

In his classic work on manias, Charles Kindleberger maintains that a regulatory backlash seeking to make amends for the recent excesses always follows on the heels of fervor. In the earliest days of introspection, Wall Street analysts looked to be the regulatory scapegoats. The reasoning was simple: they had the information, touted the stocks, had everything to gain from their sale, and gave the public no warning of the rough time ahead. The logic seemed flawless until the basic premises came to be examined later in the year. Did they indeed have the information? Did the "pro forma," "excluding onetime charges," "pooled interest" accounts bear any resemblance to what a company, particularly one as circumspect as Lucent, earned? Once Enron became a household name it became clear that analysts, like the masses they preached to, might have been woefully short on information themselves. Sidestepping the question of whether they should have known better and probed more deeply—after all, these are the people who are paid millions to read the mind-numbing SEC filings and have the "not for attribution" communications channels with management—had accounting standards sunk to a level where the information provided was largely useless? Kindleberger's study of the past points out the predictability of the accounting frauds that characterize the final throes of a mania. The behavior followed a predictable pattern: "At a late stage, speculation tends to detach itself from really valuable objects and turn to delusive ones. A larger and larger group of people seeks to become rich without a real understanding of the processes involved. Not surprisingly, swindlers and catchpenny schemes flourish." It soon became clear that the telecommunications industry was riddled with the expected swindlers and that it would take years to unravel the accounting frauds perpetrated during the final years of the boom. As accounting fraud plunged MCI/WorldCom into bankruptcy, former Baby Bell Qwest had to restate $1.2 billion in improperly realized revenues, and Lucent and later Nortel had their accounting practices investigated, no one could doubt that the boom had wrought the usual criminal elements. At ninety-one years of age, Kindleberger remarked wistfully that he would like to be young enough to add a chapter on Enron to his classic work. But he need not have worried about the lasting value of his study; only the names had changed.

In February 2001, the Securities and Exchange Commission's enforcement division began a formal investigation into Lucent's accounting practices during 2000. The investigation was triggered by Lucent's own restatement of revenues in the final quarter of the calendar year and its own contact with the SEC to apprise the regulators of the problems. The SEC probe was to look into each of the deals that needed to be restated, including inventory sent to distributors that was never sold to final customers, software licensing agreements, and onetime discounts, called "nonrecurring credits." In 2004 the SEC complaint alleged that the overstated earnings were due to improprieties by a number of individuals, but also that Lucent's internal controls were deficient.

Perhaps the most damning criticism of Lucent was that it failed to cooperate with the SEC and actually hampered the investigation. For this the SEC revised Lucent's original agreement in principle and fined the company $25 million.

Lucent's greatest threat in 2001 came not from its difficulties with products, people, strategy, or even execution, but from the less exciting and far more dangerous balance sheet. The massive swing in profitability meant that Lucent was burning cash like kindling on a campfire. This was not a problem that management could work out over the course of the year by refining its strategy, but rather a life-and-death issue. In January, Lucent had sought a short-term line of credit from its bankers and been forced to pledge assets as collateral. It was an extremely negative sign from lenders who had once been eager to do business with Lucent. It was highly unusual for investment-grade-rated companies to need to pledge security for short-term money. The implication was that, after two demotions in as many months, Lucent's continued existence as an investment-grade company was anything but assured. Any bonds would need to be issued in the junk credit market, and money market funds cannot buy commercial paper from companies that are under review for a downgrade; thus another avenue for funding the company's operations was vanishing.

In December 2000, Lucent's credit rating was downgraded by Standard & Poor's for the first time. At this point Lucent still maintained a BBB+ rating, which indicated that it had adequate resources to meet its financial commitments. Over the course of 2001 and into 2002 there would be six more credit downgrades by Standard & Poor's. On June 12 Lucent was downgraded to a junk credit rating. The significance of this was that many fund managers were not permitted to hold below-investment-grade bonds.

Lucent had stepped onto a downward spiral. As its credit ratings were slashed, borrowing costs rose. As the costs rose, the amount needed to borrow also increased. Without new funding commitments, Lucent would run out of cash. Schacht's attentions were consumed with extricating Lucent from this emergency. Everything else, including the search for a permanent CEO, was put on the back burner while the cash crisis was addressed. If Schacht did not immediately strengthen Lucent's financial positions, all his other efforts to right the company would be in vain.

The last thing any company needs to do during a week when it is facing a weakened sales picture, an SEC investigation, and a debt rating that has just been lowered to one rung above junk is to go to its bankers and ask for more cash. Yet one of Lucent's existing $2 billion credit lines was set to expire on February 22, and Hopkins faced her most difficult challenge at Lucent. Although its credit rating had been lowered just days earlier, the rating agencies were explicit in advance that if Lucent was not successful in obtaining this credit, it would be reduced to a junk bond credit. Doing Lucent's banking business had once been a glory that banks had reveled in, but this time when Hopkins came to call, looking to renew $2 billion and make another $4.5 billion loan arrangement, no one was interested in picking up the telephone.

Days before Lucent's loan was due to expire and with no firm commitments in hand, Schacht made his final appeal on a conference call with thirty bankers. His pitch was one part nostalgia, one part patriotism; as he told his bankers, "Our team is working around the clock to restore an American icon." Schacht emphasized Lucent's strong points: the restructuring plans were already in place, and he believed that he had the management team to carry them out. The alternative was clear. If the bankers were not forthcoming, Lucent would be forced to draw down an existing short-term credit line due to expire in a few days. That credit had been extended before Lucent's most recent credit downgrade and, while not sufficient to meet all of the company's needs, the money came with a lower price tag. Schacht's message was unmistakable: give us more money now for which we will pay you a higher rate [more financing for a longer term], or we start drawing down the ready cash. "We don't want to do this," Schacht told his audience, "but we are running out of time. We have gone out of our way to put a responsible deal in front of you."

Hopkins played the bad cop to Schacht's good cop. She preyed on the commercial banks' weakness. All commercial banks want to grow up to be investment banks, and since the repeal of the regulations once separating them, they have had reason to believe that such a transition is possible. Hopkins made her position clear. Only those institutions that could find their way to lending Lucent funds now would be considered for fee-based investment banking business later. The threat to withdraw Lucent's investment banking businesses, with thirty-eight acquisitions, two major spin-offs, and countless sales and financings, was not to be taken lightly. The Agere IPO had yet to be completed and, although the value of the deal was rapidly declining, it still appeared to be worth $7.5 billion, perhaps the second-largest IPO ever. The fees from a deal of this size would run into the hundreds of millions. Spurning one of the country's largest companies did not seem like the path to investment bankinghood for any of the commercial bankers listening to Hopkins that day. "This is the most extreme example we've seen so far of commercial banks using their lending muscle to displace traditional investment banks as underwriters," said Samuel Hayes, professor of finance at the Harvard Business School, referring to the Lucent loans. "It is pay to play."

Schacht was persuasive, and those who heard him responded. To help swing the deal, he promised to stay on as CEO until the crisis had passed. Mellon Bank jumped on board during the call, indicating interest and turning the momentum in Lucent's direction. Before the five o'clock deadline, more than a dozen reluctant bankers had agreed to Schacht's terms. Lucent looked as if it was out of the credit-rating woods. The new financing gave the company some breathing room, and Schacht argued that it would help put Lucent back in a profitable position.

The refinancing came down to the wire. So tight was the timing that Hopkins did not attend Lucent's annual meeting in Florida but was instead busy arranging loan agreements up until 11:00 P.M. on the night of the deadline. The conditions to which Lucent submitted included using shares of Agere as collateral. The bankers allowed Lucent to proceed with the IPO, but the company was precluded from doing the spin-off, slated for September 2001, until it had replaced this collateral with $2 billion of new nonoperating capital. These new monies needed to be raised from either further asset sales or new financings. It was a tight financial bind and would consume top management's efforts for the year. If it did not successfully spin off Agere, Lucent would continue to be the majority shareholder, and this would cause the stock price to crater. Nonetheless, raising $2 billion did not seem like an excessively high hurdle, as there were still many valuable assets that could be shed. Lucent owned the second-largest optical fiber business, after Corning, and raging demand for the glass strands was projected to continue. Corning, Alcatel, JDS Uniphase, and Pirelli SpA all had showed interest in the division. If Corning bought the business it would control 75 percent of the fiber market; the position of the two major suppliers was that dominant.

Raising money was proving more difficult by the day. The credit spreads on Lucent's thirty-year bonds widened out to nearly 4.50 percent above the yields on similar United States Treasury bonds, from a more usual 1.30 percentage points, indicating how costly capital had become for the company. In a year of increasing gloom, Agere was held out as a bright spot, a still-valuable jewel that could be converted to much-needed cash. As it turned out, market conditions were such that raising $2 billion was far more difficult than Lucent had anticipated and the spin-off of Agere would be delayed another ten months.

Agere was to be severed from Lucent in the same way Lucent had been separated from AT&T—in two stages, with an IPO in the spring of 2001, preceding the eventual spin-off on September 30. Again in parallel, Lucent was transferring $2.5 billion of debt to Agere (about one-third of its total debt to a company worth one-tenth of its revenues), not in the hope of being re-rated but in a vain effort to avoid being branded with junk bond status. However, two factors complicated this process. The value of Agere fell sharply between the time McGinn had first indicated Lucent's intention to sell in July 2000 and February 2001, when the underwriters began to price the deal. At the time of McGinn's announcement, the division was worth perhaps $50 billion to $75 billion, although some speculated figures as much as double that. What had once looked like a smart strategic move had begun to look like a desperate measure. When Lucent formally announced the deal on February 7, the price of the IPO stood at between $15 and $20 per share, suggesting that the company's value had fallen to roughly $25 billion. Then came another slide. By February 20, Lucent was indicating $16 to $19 a share, and six days later $12 to $14. The offering was delayed and while at first he suggested that this was simply a minor event, Ben Verwaayen, co–chief operating officer, later announced that Lucent was considering canceling the offering entirely and would make a decision over the following week. The credit agencies made it clear immediately that they were not giving Lucent that option. If the proceeds from the offering were "materially lower than anticipated" or the date was delayed beyond March 31, Lucent's credit rating would be reviewed again. But Lucent and its bankers could not round up enough investors interested in paying $12 to $14 a share for Agere as they watched the stock prices of the new company's competitors wane.

Selling Agere dirt cheap was a desperate act and the timing could not have been worse, but Lucent was running out of options. In better circumstances, Lucent would have withdrawn the IPO and waited for a more hospitable climate to sell off its largest asset. Like Lucent, Agere was watching demand for its products vaporize. Nortel had issued a second earnings warning for the same quarter (as conditions deteriorated faster than it had expected) on March 27, 2001, the day before Agere filed, and the new company was forced to admit in its IPO filing that it might be facing a "significant operating loss." However, since the stock had already been used as collateral, Agere had to be monetized. In the end, Lucent got only $6 a share for Agere, netting $3.6 billion, on March

28. Agere was the sixth-largest IPO ever, just ahead of Lucent's seventh place. As if to firmly signal the end of an era, a high-tech IPO with a blue-blood pedigree, the world's largest maker of optoelectronic components, closed its first day of trading only two cents above the flotation price.

On April 4, the fifth anniversary of Lucent's IPO, rumors of the company's imminent bankruptcy swirled through the financial markets. They were so persuasive that Hopkins was forced to make a public statement insisting that the gossip had no credence and that the speculation of Chapter 11 was "baseless and irresponsible." Lucent was not facing bankruptcy or anything close to it, but it was a sobering experience and a testament to the speed of the times that fifteen months after being feted as the next Microsoft or IBM, Lucent was fighting off rumors of its demise. The share price had collapsed by more than 90 percent from where it had been when McGinn had insisted that there was no systemic problem, just difficulties in execution that would be straightened out soon. It was an extraordinary round trip. Over the course of exactly five years, the stock had gone from a split-adjusted price of $5.70 to $59.36 and back to $5.70.

Rumors of the company's insolvency were only part of Hopkins's difficulties. She was not a product of AT&T and despite a year at Lucent had not absorbed any of its low-key, understated ways. A number of Lucent's executives who worked with her found her style abrasive and her absorption with her own career troubling. She, in turn, found a rigid culture, hostile to newcomers and new practices. Opinions about Hopkins are sharply divided. More than one board member has said she came to the board and explained in detail Lucent's financial control problems and her strategy for setting things right. She brought with her a strong financial team to execute the changes. Some of those who knew her at Lucent said that she went to great lengths to distance herself from past practices, often speaking badly of action taken before her tenure. She was candid about the company's financial problems and exhibited little patience for the myriad of excuses for why things had gone wrong. She put in place systems for more closely monitoring Lucent's profitability by region, product, and customer that threw some of the company's weaknesses into sharp relief. Her criticisms hit right at the veteran staff. While this was perhaps not politic, it may have been fair and accurate, though painful for those there at the time to hear. What Hopkins saw as direct and honest, her new colleagues saw as abrasive. She had not decoded the company's unwritten rules, and as one executive who worked closely with her and praised her efforts remarked, "There was an organ rejection."

Hopkins had come to Lucent because she shared a vision with McGinn of what they might build together; with Schacht she shared none of this. Hopkins described herself as a "relentlessly executing, no-holds-barred, take-no-prisoners, be-in-front-with-the-shield-and-sword, go-to-war, stay-in-the-ditch-with-you executive." While Lucent's executives were highly motivated and committed, there was a vast difference in style. Schacht is the picture of modesty and understatement, beginning most sentences with "we," not the royal but the inclusive, and liberally redirecting credit away from himself.

Hopkins knew that her relationship with Schacht was not what it had been with McGinn, and on the first Saturday in May 2001 she requested a meeting to discuss matters going forward. The board always knew that Hopkins was interested in the CEO position and that McGinn had told her that at the right time she would be considered for the spot. She had no such deal with Schacht, who within a few months would need to look for his own replacement. The two executives did not agree on how to proceed. Schacht favored the more gradual approach, cutting only where absolutely needed. Hopkins saw a bleaker picture for Lucent and the industry and wanted to make deeper cuts more quickly. Hopkins had already incurred Schacht's ire by criticizing his plans to merge with Alcatel. She favored either reengineering the company as a stand-alone entity or dismantling it by selling off the pieces. Schacht was obviously of a different mind, not viewing Hopkins as a team player committed to his plans. There was little overlap in their visions of the company's future, and Schacht asked her to leave.

Publicly, Lucent praised Hopkins, calling her "instrumental in helping instill more financial discipline," but privately some executives questioned her achievements and ability to read the culture's unwritten rules. Hopkins was gagged by a severance agreement and refused to speak with the press after her departure, yet she made it clear that she had found the situation at Lucent an enormous disappointment. She had had no idea of the depth of Lucent's problems before her arrival and while insisting throughout her tenure that she played the cards she was dealt, when asked by *The Wall Street Journal* about wanting to leave, she admitted, "I would be disingenuous if I said the thought didn't cross my mind many times." After Hopkins's departure the press was scathing, but with some prescience she was the one who only a year earlier had told an audience, "If you lose the game, you can expect 12 weeks of intense media scrutiny about what went wrong—and who's to blame. And that's when the media puts its phasers on 'stun.' "

Reverting to form, Schacht replaced Hopkins with Frank D'Amelio, the former group president of switching solutions and a twenty-two-year veteran of AT&T/Lucent. "In rebuilding the company, we need someone who understands both broadband and the balance sheet," a Lucent spokesperson reasoned. There can be no doubt that D'Amelio was more familiar with Lucent's businesses than Hopkins. Yet as Lucent struggled to forge a new path, it seemed to be ridding itself of all those who had ever been down one.

A WHITE KNIGHT

Merger rumors had swirled around Lucent throughout the entire life of the company, and they intensified as its troubles deepened. In the spring, Verwaayen spoke with Nokia and Siemens to see if they had interest in some sort of connection, but the conversations were vague and short. Lucent had dialogues with Marconi, British Telecom, and Motorola, but Alcatel, the French communications equipment supplier, looked like its best prospect. Usually merger negotiations take place in secret, and headline news stories are the first thing many hear about a deal that is already a foregone conclusion. By conducting business in this way CEOs save themselves and their organizations the external pressure that comes with a barrage of media commentary and leaks, and they minimize their employees' anguish over the unknown outcome. None of these conditions prevailed as Lucent and Alcatel endured a highly public, overanalyzed series of negotiations that were drawn out for much of the spring of 2001. Lucent was negotiating from a position of weakness. Schacht had not been able to stanch the flow of losses, and whether there was true panic or not, the discussions with Alcatel had the appearance of a last-ditch effort.

Conversations between the two companies began in March, when Lucent tried to sell its fiber-optic cable unit to Alcatel. The cable business was one that could easily be split off and sold in the ongoing fire sale. Serge Tchuruk, Alcatel's CEO, announced that his company had made an initial offer for this business, but at the same time he began inquiries about a much more extensive deal. In their earliest discussions, there was talk of the bid for Lucent being at 20 percent above the market price of $33 billion, with the fiber business making up about $5 billion of the total. The conversation stalled the sale of the fiber unit for the duration of the negotiations while the price continued to fall. A number of potential bidders for the fiber-optic business, among them General Electric and Tyco, lost interest.

While both companies had been badly weakened in the previous nine months, the combined entity was seen to be a formidable competitor, something more than the sum of its parts. Tchuruk was considered a turnaround expert, something Lucent sorely needed. Those critical of the deal suggested that the two would simply be combining their problems and that the difficulties in integrating the operations would swamp some of the cost savings. For months, management from both companies kept in close contact, with O'Shea meeting with the Alcatel COO at least once a week to coordinate and prepare for full merger talks between the respective CEOs. The risks in such a deal were enormous. Large international mergers have a mixed history of success. Large technology mergers have an even worse track record, of which Schacht was all too aware. Thus the notion of integrating two troubled technology companies from vastly different cultural footings would be a fantastically difficult feat to pull off. While the combined company would be the largest telecommunications equipment supplier in the world and would probably find itself in an improved competitive position, the execution of the merger would potentially be fraught with difficulties.

In May, both boards agreed to let their managements proceed with discussions. At a sixteenth-century French castle (fitted with a single telephone line) that bore little resemblance to the offices at Murray Hill,

Lucent's top executives and their legions of bankers and lawyers met for three intensive days, with discussions of the merger occupying breakfast, lunch, and dinner. Tchuruk spoke passionately about the opportunity to combine the best of both companies. The deal as it was negotiated was a stock swap that gave Lucent's investors Alcatel shares with no premium on the price of their shares and valued the entire company at $23.5 billion. "This is not about short-term premium, but long-term opportunity," the Lucent side continually reminded the press. The new company would be incorporated in France and headquartered in New Jersey, and Alcatel's shareholders would own about 58 percent of the combined entity. But Lucent was married to the notion that this was to be a merger of equals. One line of thinking was that Tchuruk would be chairman and Schacht vice chairman. It becomes slightly difficult to see how this would have been a merger of equals, but if Lucent had equal board representation, its influence would remain strong. The chairmanship, company name, and board composition remained bargaining chips, unsettled until the last.

On Sunday, May 27, 2001, the teams reconvened in Paris. There they reviewed the final merger documents, press releases, and the information to be given to Wall Street, and the Americans flew home, with only the composition of the board of directors left still unresolved. Conference rooms were booked at the St. Regis Hotel in New York for an analysts meeting, and a press conference was planned for Wednesday to reveal the details of the merger. The deal was so close to being complete that the two companies were arguing about the placement of their respective logos behind the podium. Fitzgerald had begun to prepare notes for the executives to use at the press conference, but she had left a huge blank under the section entitled "Board of Directors." Schacht had displayed enthusiasm for what he characterized as a merger of equals, telling the two management teams, "This company will be like the NFL draft—the best athlete available will be chosen for every position," in an allusion that might have been lost on his French counterparts.

Monday was Memorial Day in the United States, but negotiations continued between Paris and New Jersey by phone. Despite the fact that the public announcement was only thirty-six hours away, the composition of the board of directors was proving to be a sticking point. The Lucent team was losing patience. For Schacht, this was a nonnegotiable point. He made it clear that if he did not have equal representation, it would be impossible to suggest that this was anything other than a takeover by Alcatel. Senior managers at Lucent did not believe that they could sell the deal to their shareholders if it looked like a rout of the home team. Conversations between the sides had centered on a sixteen-member board, with Lucent holding six of those seats, Alcatel filling eight, and the final two to be selected at a later date. Schacht wanted to choose who would fill the two empty spaces, and Tchuruk wanted to have a say in the appointment of at least one. Finally Lucent gave up, saying that Alcatel had so altered the deal in the final stages that it had no choice but to walk away.

During their stay in France, the Lucent team had a good look at Alcatel's finances and were cautioned by what they saw. Alcatel's shares had held up much better than their own, but this new, not-yet-public information indicated that this might not continue to be the case. The inability to agree on the composition of the board proved to be the ultimate deal breaker, and a press release was issued saying that the merger was dead. In its version of the press release Alcatel was also forced to issue a profit warning, coming clean for the first time about how fast its business was deteriorating. The Lucent team was appalled, as it realized that any announcement of a merged company would have had to contain a profit warning as well.

A number of Lucent's high-ranking executives had been opposed to the merger with Alcatel. Some felt on emotional grounds that Lucent could make it on its own and that it had a heritage and employees to protect. One senior executive conceded that a merger might be necessary at some point but noted that it was crucial for Lucent to get its house in order and strengthen its bargaining stance before sitting down to negotiate. Lucent had struggled badly to integrate Ascend into its operation, and there is no evidence that a merger with a foreign company riddled with overlapping product lines and, as it would turn out, its own financial difficulties would have been any less challenging.

There were some last-minute efforts to bring the deal back to life, and the fact that the two CEOs parted amicably was viewed as a sign that options were being kept open. Now Lucent needed to face up to its own problems. The refinancing, asset sales, and merger talks had delayed the restructuring. Lucent fell behind on its schedule for cost cutting and during the talks postponed some of the intended asset sales. Prices for those assets, in particular the optical fiber business, had subsequently plummeted. Schacht was confident of Lucent's progress in its own restructuring plans. "Why would we walk away from a deal unless we were confident about going it alone?" he asked. "That's not a sign of weakness, it's a sign of strength."

The bond market did not see it that way. Alcatel had a higher credit rating than Lucent and Lucent's bonds rose in price during the talks, offering the possibility that Lucent could be re-rated in a merger. Within minutes of the announcement that the deal would not materialize, Lucent's bonds lost ground, dropping from $.95 on the dollar to $.86. The bond market had an accurate reading of the situation, and the following week its fears were confirmed when Lucent slipped farther down the credit spiral and was downgraded to junk bond status. One Lucent senior executive looked back on the deal with Alcatel as a missed opportunity in an industry inevitably headed toward integration. He says now, "Alcatel would have been good. However difficult, the integration would have been executed. There was a good geographic fit, and with a much larger revenue line, Lucent would have been the largest telecommunications company in the world."

After the Alcatel deal receded, Lucent still needed to sell its fiber business. The delay had been costly, and the sale of this unit followed the trend of other deals, starting with optimistic projections of $8 billion to $10 billion, slipping soon to $5 billion to $6 billion, and ending up with a single interested bidder, Furukawa Electric, for $2.3 billion, in a deal where the price was negotiated down twice. The negotiations took most of the year, and the price only declined as bidders faded away. Schacht tried to put a good spin on the situation, claiming that Lucent was under no pressure to sell and would get a good price. But Lucent's pursuit of cash in an environment with collapsing values left the company hostage to a market in which buyers were quickly evaporating.

If the telecommunications industry was creaking in 2000–01, it collapsed in 2002. Most of the CLECs were either gone or mired in bankruptcy, and the Baby Bells, disabled by five years of overinvesting and price-cutting, halted their expenditure on equipment. Lucent's executives, like most of its competitors', had been surprised at the industry's growth and were now entirely unprepared for its decline. Each time Lucent's customers conveyed their spending plans they quickly recanted, facing a bleaker marketplace than they had expected, and in desperate need of shoring up their own balance sheets they began to hoard their cash.

Despite continued promises about a return to profitability, Lucent never stabilized in 2001. Each quarter's forecasts were quickly proven to be hopelessly optimistic, as the market for telecommunications equipment evaporated. The credit-rating agencies continued to lower Lucent's rating, citing the company's inability to cut costs fast enough to become profitable. Each quarter, Lucent's management predicted that the bottom was in sight, only to discover that the hole was far deeper than it had thought. Despite continuing to shrink the company, at no point did management make the drastic cuts that would reduce costs to below the revenue level. Lucent closed 2001 with losses of $16.1 billion. This loss wiped out all of the company's profits throughout its history. All totaled, Lucent has never made any money. The company's profits in the early years were quickly erased by ten consecutive money-losing quarters. As the company's stock price slipped below a dollar and its bonds traded at twenty-nine cents, a price that implies default, the talk of bankruptcy was out in the open.

Lucent was whittled down to what had once been Network Systems. Bell Labs had shrunk beyond recognition from the research arm of a company supporting a million employees to one supporting fewer than forty thousand. The company's revenues for 2002 were $12.2 billion, levels last seen in the 1980s. And the management was now, after two years of trailing the decline, preparing for worse. In late 2002, Lucent promised to reduce its cost structure to the point where it would break even on revenues of $10 billion a year, what had once been a healthy quarter's sales.

CHAPTER TEN 2002–2003

2002–2003 Stock price   High: $6.13   Low: $.55

> *This was once a great company, and it was*
> *something to be proud of. I personally don't*
> *want it to go out on a low note and I don't think*
> *any of us want Lucent's last days to be down.*
>
> Janet Davidson
> president of Lucent's Integrated Network Solutions

Despite insisting that his role as CEO was temporary, more than a year after his return Schacht had found no replacement. While head-hunters were busy sifting though résumés and speaking to candidates, the number-one internal candidate, Ben Verwaayen, left to become CEO of British Telecom. Dan Stanzione, the retired head of Bell Labs, was also considered. Some Lucent insiders believe that both men would have made more radical changes to Schacht's plans and to the existing executive lineup. Stanzione, they suggest, would not agree to take the job with Schacht looking over his shoulder. No one knew the industry, the engineering, or the company better than he did, and he hardly needed to be mentored or second-guessed. PR head Kathy Fitzgerald reflects the sentiments of many who worked with Stanzione in saying,

"I'm not surprised that Dan was spoken of so highly by everyone who passed through Lucent. The original Lucent team was a virtual 'dream team' of top management talent with very smart, knowledge-able, articulate, competitive, creative leaders who were energized by the opportunity to create a new company. But even in that company Dan was special. He was a thoughtful man who listened. He was a results-oriented manager who inspired. And everyone—at every level—felt that they could trust Dan. He was in it for the whole team, not for himself."

Potential CEOs were not beating down Lucent's door for the opportunity to turn Lucent around, and despite the fact that over one hundred candidates were considered, the search lengthened but no interested candidate was deemed to be suitable. Dan Plunkett, Lucent's longtime consultant, remembers the search: "Henry needed a different kind of CEO for Lucent in 2001 compared to 1996. Lucent was in a tailspin in 2001, and he needed a leader who was both a technologist with deep industry knowledge and a strong operational leader who could lead the turnaround. He searched for such a leader and all the usual suspects turned up, but none of them fit the bill."

In January 2002, Schacht announced that Pat Russo would return from her position as president of Kodak to Lucent as the new CEO while he continued as chairman. Russo was a favored choice among many of the Baby Bell CEOs, who knew her well, but a disappointment to those who hoped that Lucent would seek out new talent. "Without fresh blood at the senior management level, Lucent runs the long-term risk that usually accompanies inbreeding," said Steven Levy. "In other words, stagnation and the lack of innovation." Russo's plan, as she stated it, was to carry on with Schacht's plans, cutting costs and seeking out what little business Lucent's traditional customers afforded. Two executives who know both Schacht and Russo well have argued that Schacht felt very bad about the way Russo had been treated when she left Lucent and that he believed that her strengths in dealing with customers were what the times demanded. She was certain that she was up to the challenge and on one of her first days back remarked, "This is a job I feel like I have been preparing for all my life."

For the third time, Schacht had passed up the opportunity to bring new leadership to Lucent. First when establishing the company, later when he returned to the helm, and finally when selecting his successor, he insisted that Lucent did not need a new face. In bringing in Russo, he maintained the status quo. She was part of the old fabric of Lucent, invested heavily in its past, and in her first year in charge did little that was a break with this heritage in terms of either personnel or strategy. No one could have been surprised at Schacht's selection of Russo. In his first week back as CEO he made it clear that he would not rush the process, saying, "When you go outside [seek a CEO from outside Lucent], you always have to be super careful, because it is the devil you don't know." Like McGinn before her, Russo asked Schacht to stay on as chairman, and executives close to them suggest that she consulted him closely on major decisions. He relinquished this role at the February 2003 annual shareholders meeting, while still remaining in the positions of senior adviser and company director.

As Schacht's role within Lucent faded, Russo took more aggressive action in remaking the company's management. She appointed new directors to the board and eased out a few of the long-serving executives. Customers speak about Russo in almost reverential tones. "You walk away knowing that she is genuine and really cares about people. She's interested in your success as well as in the company's success," says Harry Carr, onetime Yurie president and now CEO of a start-up. "The best chance they have is with her at the helm." "Since she has been there, they're listening to their customers," says Paul Lacouture, president of the Network Services Group at Verizon. "Whatever she tells you she'll get done, she does. They've become a stronger supplier—a more flexible company." But Russo knows exactly how close Lucent came to going over the edge of bankruptcy and that until the telecommunications market has a renaissance, nothing is assured. She is realistic about Lucent's troubles and assured in her approach. Taking a line from Winston Churchill, she says, "Never, never, never give up. I think at some point, every effective leader facing a tough time, a tough market, a tough economy, has to reach deep down inside and say, 'Never give up!'"

One of Schacht's final tasks as CEO was to resolve the two major lawsuits that still hung over Lucent. He had vowed that he would never settle the Aversano lawsuit. To do so would be to admit that she had been a whistle-blower and that Lucent had ignored her warnings. Schacht did not believe this to be true. He was convinced that she had continued to give McGinn assurances that North American sales would reach revenue targets that they never attained. Schacht told colleagues that for him it was a matter of principle, and he did not want the company's shareholders, its employees, or the SEC to believe that any of her charges had merit. Lucent's lawyers told him they were more than happy to try the case; their lead counsel even joked that Schacht, with his statesmanlike appearance and demeanor, would look great on the stand. Just two weeks before the case was due to go to trial in January 2003 as part of a settlement, Aversano dropped the charges that she had been a whistle-blower, and Lucent paid her severance. But Lucent had suffered greatly because of this lawsuit. The piles of confidential information that were released into the public domain—internal sales records, faxes, e-mails—shed an unflattering light on the company.

In 2004 Lucent also settled its problems with the SEC in an agreement that would end a lengthy investigation. The company did not admit to any fraud and consented not to commit any fraud in the future. Later, Lucent faced a fine of $25 million because, according to the SEC, it wanted to make clear that "companies whose actions delay, hinder or undermine SEC investigations will not succeed." Three Lucent employees settled with the SEC and paid fines without admitting or denying the allegations against them.

The case against the six other Lucent employees (and one from Winstar) will now proceed to the courts. Despite admitting no wrongdoing, Schacht's deposition makes it clear that Lucent's sales practices in the 1999–2000 period had strayed some distance from the kind of behavior that he and investors would have expected. In her lawsuit against Lucent, Aversano's lawyers had grilled Schacht: "Are you aware of any public disclosure that Lucent made, in fiscal 2000, regarding the extent of selling forward at discounts?" "Not that I am aware of," Schacht replied. Pull-ups? Gambling on pulling current business forward? Gap filling? Schacht's answer was always the same. As he readily admitted that all of these practices and more had become evident to him in a matter of two weeks after his return, the question hung in the air, too speculative for any lawyer to allow in a deposition: How were investors to know?

In winding down some of its legal difficulties, Lucent also settled a consolidated class action suit that had been filed three years earlier. In the suit, investors claimed they were misled by Lucent's rosy forecasts and had management been more forthcoming about its optical misses and its accounting been more reflective of the company's economic reality, many would not have held on, waiting for the stock to rebound. In settling, the company again did not admit wrongdoing, yet it agreed to pay out $315 million in common stock, cash, or both. The company would take a charge of $420 million because of the settlement. The plaintiffs' lawyers had initially hoped to get a great deal more. When they began the lawsuit, Lucent's shares were trading at six dollars; during the course of the case they dropped to sixty cents. At that point Lucent was worth only a few billion dollars despite the fact that the aggregate claims in the class action suits totaled over $60 billion. And although the class action settlement is the second largest ever, both sides understood that the cost to Lucent of losing even one of the fifty-four cases filed against it made the prospects of bankruptcy real.

Three years after its downfall, Lucent's problems were far from over. Retrenchment had failed to secure a profit, and despite two years, under Schacht and then Russo, of pledging a return to profitability, Lucent was still issuing untimely warnings of loss-making quarters well into 2003. The industry had not rebounded, and promises to shareholders had not been kept. Globally, there was still far too much capacity and duplication in telecommunications manufacturing and research. If history repeated itself, there would be more mergers and further consolidations. The opportunities that Lucent had walked away from, a tie-up with Cisco or Juniper, a merger with Alcatel, along with half a dozen more casual overtures, might all need to be revisited. As Russo looked back on her first eighteen months at the helm in June 2003, she seemed more realistic about Lucent's prospects given the decimation of telecommunications, saying, "The rates of decline in spending have decelerated, and therefore the industry feels like it's moving more toward stability. I would not call it a bottom, and I would not call it a recovery. We're seeing an improved condition over what we saw in 2002, although we are not yet into absolute stability on the way to recovery. So it's better, but still not good enough." But even as Lucent was forced to issue yet another earnings warning in July 2003, giving investors worse news than they had predicted, with a sharp decline in revenues, Russo sounded exactly like her predecessors when she made the case that the bad news was just a "hiccup." At no point over the four years of Lucent's decline did any executive offer a dismal forecast or resist the urge to tell a story of how things would look up. Despite the unending damage to their credibility and what must have from inside the industry looked like an imploding market, investors were not fed the unvarnished truth.

HALF A CENTURY GONE IN A MATTER OF MONTHS

If investors were disappointed over Lucent's travails, employees were devastated. Wall Street traders and investors thrive on the wild gyrations of the stock market; for them they are a source of opportunity as well as losses. But for those who toiled at the company, the human cost of Lucent's nosedive fell hard on workers and managers in plants and laboratories all over the United States. And it all happened so fast.

Western Electric opened Merrimack Valley Works in 1953 on what had formerly been a vast tract of farmland. The mammoth plant, which covered more than two million square feet, was located thirty miles north of Boston on 157 acres at 1600 Osgood Street, North Andover, Massachusetts. One labor union executive, whose parents both worked at the plant, recalls that on his first day he was lost for more than an hour as he wandered the massive complex in search of his post. In time it would be Lucent's largest facility, so large that employees could exercise during lunch by speed walking three to four miles within the corridors without ever leaving the building. MVW even had an Emergency Response Team with its own miniature fire truck for use inside the building. "It was a facility that was all-encompassing," said Brian Major, a senior technical manager and local selectman who worked at the plant for fourteen years. "You had your own police; you had your own fire and rescue, your own hospital, your own library located in the facility and stores for buying safety glasses and shoes. You name it, you could do it in the facility. The only thing you couldn't do was dry-cleaning."

When Western Electric was hiring it recruited from the families of its existing employees, and AT&T had done much the same. Extended families spanning the generations worked at MVW; it was a way of life for an entire community. The site could accommodate the more than ten thousand workers who filled it in the 1970s, but by the time Lucent came into being, the numbers had dwindled by half. "In the early 1980s, you needed 500 people to make copper coils for phones," Joseph Kanan, president of Local 1365 of the Communications Workers of America, said. "By the late 1980s, you only needed 50 people because of automation." Yet even with fewer employees, life was good at MVW. Lucent invested roughly $300 million to update and expand the plant and added more than seven hundred jobs, bringing the total to more than 5,600 employees.

In Merrimack Valley, Lucent was producing the world's most advanced optical products, among them the Lambda router, and paying premium salaries to attract talent in the competitive Boston marketplace. "Merrimack Valley is the premier optical networking manufacturing site in the world and continues to demonstrate its operational excellence through its talented workforce," plant manager J. R. Newland said in 2000. "We will continue to recruit locally, across the country, even worldwide, for the best talent available." In 2000, Newland and Lucent announced huge plans for MVW; it would be the company's cutting edge, a plant that would produce products as soon as Bell Labs spit them out.

"People would kill to get a job at Lucent," recalls one union official in MVW, because Lucent set the wage standards for the area. Month after month the plant broke its own earning records as Lucent ran the operation three shifts a day, seven days a week. The company was awash with overtime pay, and compensation for many ran into six figures. But wages and steady work were only part of the package. Gary Nilsson worked at Lucent his entire career; his mother had worked at MVW before him. When Newland called the first all-employee meeting at MVW in month, year, Nilsson was surprised. Nothing like it had been held in his tenure. He recalled, "Our plant manager told us, in captive meetings, that Lucent was going to make us all rich, rich, rich. He advised us to take everything we had and invest it in Lucent stock. Unfortunately, a lot of our members took his word for it. After all, he was a corporate executive with the thousand dollar suits; he must know what he's talking about. For a while everyone watched their savings plans explode. Everyone was getting high on the golden ride; no one stopped to look around. They were getting rich and that's all that mattered."

When times had been good, Lucent had outsourced some of its MVW circuit package manufacturing. The stock market had cheered McGinn when he had announced the change, as it meant more work done at cheaper wages. At first employees at MVW had not been overly concerned, because Lucent clearly had more work than it could handle and assurances were given that no job losses would result from the outsourcing. In January 2001, Lucent announced that it would close plants in Columbus, Ohio, and Oklahoma City. Workers at MVW thought the bad news might be good news. There had been less work at the factory since the summer of 2000, but now perhaps things would change. "We thought that after the other plants closed the work would come to us," explains Maddie Carrier, who worked for Lucent/AT&T for twenty-six years. (Her husband had put in forty-one years, and thirty-eight other relatives made their living there.) But that was not to be.

Over the course of 2000, as Lucent's stock price slipped, faith in its shares began to teeter. "Our plant manager held another captive meeting to tell us not to worry; he had the charts to prove it," Nilsson recalled. " 'Just keeping putting everything into Lucent.' The truth was, Lucent was putting it to us. The stock crashed. Overnight thousands of our members lost their entire life savings. We had members who couldn't retire because they lost everything."

But for employees, the stock price was only one facet of a growing problem. As Lucent lost market share in optical over the course of 2000, the pace around the factory slackened. Employees stood around killing time, knitting, doing needlework, or just wandering the halls. And while the plant was still open all the time, not everyone was working all the time. This had often happened under AT&T; sometimes the situation would go on for months before work picked up again. Everyone knew that telecom was a cyclical business, but no one had to tell these former Ma Bell employees that life under Lucent was different.

In April 2001 the bad news came. MVW would not be spared, and layoffs were imminent. Lucent offered $40,000 severance and asked for volunteers. A few hundred people took the package, a few hundred more were fired, and everyone else breathed a sigh of relief. "We thought that was behind us, that was the end of it, although there were still people sitting around doing nothing," Carrier said. As the telecom market faltered, opinions about the seemingly innocuous outsourcing soon changed. "When the layoffs first started, Lucent pretty much avoided MVW," Nilsson said. "We had lots of workers sitting around entire shifts doing nothing; there were no orders to fill. But . . . around us [factories] were still working, building our circuit packs. Low-wage, no-benefits, non-union [factories] were doing our work while we sat idle waiting for the first ax to fall."

In July the same package was offered again and only thirty-six people volunteered to take it, although 275 were eventually laid off. At MVW it was an astonishing, almost unbelievable, turn of events. In mid-2000 MVW had been named a "Lucent Center of Excellence" and told by senior management that its business would be expanding as fast as they could find qualified staff. Now, less than a year later, these same workers, many of them second- or third-generation employees of MVW working side by side with spouses, siblings, parents, and cousins, were faced with the option of taking early retirement or hanging on to see if they would get fired.

Hope lingered all summer, as MVW employees organized and formed the Crisis in the Valley Committee with the slogan "Tell Lucent . . . Keep Family Wage Jobs in the Valley." The group staged rallies, wrote letters, and held public meetings with elected officials. When the union found out that some of the contract manufacturers would be visiting the plant, looking at the facility as prospective buyers, they held Red T-shirt Day, handing out red shirts to the workers on all three shifts. Visiting executives touring the plant saw an ocean of red shirts all bearing the Crisis Committee's slogan. Each quarter as Lucent announced its failing results, Merrimack Valley employees hoped they could hold on to a way of life. Maybe the telecom market would turn; maybe Lucent would find its footing; maybe this last round would be the final blow and the rest of their jobs might be saved.

By September 2001, reality had begun to set in. When Lucent again asked for volunteers to accept a retirement package, by now padded out with more cash and benefits, 1,420 people lined up to forfeit their jobs.

Two years later the bitterness is still palpable among those who took the packages—and those who were later fired. "They lied, they lied, they lied," says one former union official. "Each time they said it was the last time, they lied." In the end it would all be gone: the $300 million investment the massive plant, the family workplace. By the time Lucent put the plant up for sale in July 2001 the assessed value of the facility had fallen to $37 million, despite the investment and the $45 million Lucent had paid AT&T in 1996 to take possession of the facility. When Lucent finally sold the property to a developer in September 2003, the sale yielded a mere $13,860,000. "This is really tough because most of these people are right in the middle of their careers," Nilsson said. "It is one thing to lay off people who are close to retirement, or to lay off people who are real young. But these people are in their mid 40s to early 50s. They are too young to retire and they are much too old to start all over. This will really devastate these people." Most had walked through the factory doors straight from high school and were trained for nothing else. "It feels funny to leave. I came in this morning," said one employee. "I said, 'geeze.' I first came in here working in the mailroom as a 19-year-old and here I am, almost 59. I'll never go back again."

In October 2003, Lucent offered up the first bit of good news in nearly four years. While revenues were up only slightly, Lucent was able to record a small profit. Yet Russo was very careful to temper expectations when she told her employees, "Let me tell you how I define a recovery. If you look at capital spending in the aggregate—what our customers spend on things that we provide—I believe we could call it a recovery when [spending] starts to go up. It has not started to go up . . . I can't think of a customer who has yet said, 'I will spend more next year.' When that [will] happen, I don't know." Although Russo's announcement brought welcome relief from the unremitting flow of bad news, only those with the shortest of memories could believe that this was the same Lucent that had burst forth with such promise only seven years earlier.

Walking through Lucent's vast open lobby in Murray Hill, it is hard to remember that this was the headquarters of a company with nearly 160,000 employees, once the sixth-largest corporation in the United States. On the surface, things look much the same. The Lucent museum is still in place, with rows of glass-fronted cases documenting the company's twelve decades of scientific achievement. A huge digital counter tabulates the number of patents the company has received since 1925. But with 75 percent of the employees gone, the parking lot is almost empty, the cafeteria half full, and on the executive floor offices and conference rooms stand vacant, built for a much larger management team that is now off running other companies.

NOTES

In a perfect world, the more than one hundred interviewees I spoke with would have allowed me to divulge their identities. But among this helpful group some still work for Lucent, others work for companies that do business with Lucent, and still others, given the state of the telecom industry, hope to find themselves in either of these positions soon. Thus I have used the note "author interview" in cases where more information could not be revealed. Every time this vague euphemism is used, it is because the source spoke on the condition that I hide his or her identity. If I had not done this, the Lucent story could not have been told.

PREFACE

PAGE

3   *Before any important meetings:* Author interview.

3   *This fact might have triggered:* Author interview.

4   *In an unintended twist of irony, he told:* Author interview.

CHAPTER ONE

PAGE

5   *"The underpinnings of the emerging telecom":* Gretchen Morgenson, "Telecom, Tangled in Its Own Web," *New York Times,* March 24, 2002.

5   *With a collapse in market capitalization:* "Market Value Loss Has Been Unprecedented," Goldman Sachs research, November 2002.

7   *"The pace of change is escalating":* Speech by Henry Schacht, "Value-Driven Behavior Is a Business Imperative in the Global Economy," April 22, 1999, Penn State University.

CHAPTER TWO

PAGE

9   *"When everyone had a telephone then AT&T":* Stephanie Mehta, "Say Good-bye to AT&T," *Fortune,* September 18, 2001.

9   *"Changes in customer needs":* "AT&T Announces Major Restructuring for 21st Century," AT&T Press Office, September 20, 1995.

11   *"The research that was done years ago":* Paul Keegan and Katie Avoy, "Research at the Speed of Light: Lucent Technologies Recharges Bell Labs," *Upside,* February 1999.

12   *Face-to-face meetings between AT&T chairman:* Steve Coll, *The Deal of the Century, the Break-up of AT&T* (New York: Atheneum, 1986), p. 25.

13   *"Everyone knew at the time NCR was a third- or fourth-rate":* George Church, "Just Three Easy Pieces," *Time,* October 2, 1995.

15   *In any two-hour sales meeting:* Author interview.

16   *Allen told Ramqvist that a merger:* Author interview.

16   *"The market value of AT&T was being buried":* John J. Keller, "AT&T: The Second Breakup," *Wall Street Journal,* September 21, 1995, p. A1.

17   *For Lucent, these legislative:* E-mail from Kathy Fitzgerald to Lucent leadership, Re: Henry Bullet Points/*Fortune* Magazine, January 9, 2001.

17   *"And, while the rest of the world was experimenting":* Robert E. Allen, "Cutting the Barbed Wire: Lessons of a Reformed Monopolist," speech at University of Texas, October 21, 1996.

17   *As comedienne Lily Tomlin's telephone switchboard operator, Ernestine:* www.pbs.org/newshour/bb/economy/july-dec99/att_8–10.html, PBS MA Cable, August 10, 1999.

18   *Schacht thought it was a brilliant decision:* Author interview.

18   *"I'm trying to shape the future rather than react":* John J. Keller, "AT&T: The Second Breakup," *Wall Street Journal,* September 21, 1995, p. A1.

19   *CEO McGinn was the first to admit:* Brian Taptich, "Lucent Unbound," *Red Herring,* August 1, 1998.

19   *As a former executive explained:* Author interview.

CHAPTER THREE

PAGE

21   *"Either you found Rich energizing or exhausting":* Author interview.

22   *"Rich tried to turn Lucent into the company":* Author interview.

22   *Fiercely competitive, McGinn claims:* Alice LaPlante, "The Man Who Built Lucent," *Electronic Business,* December 1998.

23   *David Nadler, a consultant with very close:* Peter Elstrom, "Lucent's Ascent," *BusinessWeek,* February 8, 1999.

23   *"Being brought up in front of your peers":* Author interview.

24   *"I was and remain surprised by how quickly our people":* Jeffrey S. Young, Cisco *Unauthorized* Forum, Roseville, California, 2001, p. 120.

24   *"Rich gets people to have expectations":* Peter Elstrom, "Lucent's Ascent," *BusinessWeek,* February 8, 1999.

25   *"Stanzione was one of the smartest men":* Boxer interview.

26   *The whole process had changed:* Boxer interview.

27   *"Why can't we too grow at the same":* Author interview.

28   *Marx, who close colleagues say tried:* Author interview.

30   *"For Henry this was a huge deal":* Author interview.

30   *Allen said, "I would like you to work":* Author interview.

30   *In his press release, Allen:* Lucent Press release, "AT&T Names Top Leaders for Two Communications Companies," October 12, 1995.

30   *Perhaps the most important part of his job:* Jay A. Conger, Gretchen M. Spreitzer, and Edward E. Lawler, *The Leader's Change Handbook* (San Francisco: Jossey-Bass, 1999), p. 20.

30   *Whenever he and Schacht could not be found:* Author interview.

31   *"This isn't what I had in mind":* John Keller, "Unlikely Team: An AT&T Outsider and a Veteran Join to Run New Spin-off," *Wall Street Journal,* October 14, 1996.

31   *One senior manager remembers thinking:* Author interview.

31   *"At first we couldn't figure it out":* Schweig interview.

32   *At their initial meeting:* Author interview.

32   *Later, when McGinn recalled the situation:* John Keller, "Unlikely Team: An AT&T Outsider and a Veteran Join to Run New Spin-off," *Wall Street Journal,* October 14, 1996.

32   *From the beginning it was understood internally:* Conger, p. 20.

32   *"My advice to anybody":* Conger, p. 9.

33   *"The value system of the corporation":* Speech by Henry Schacht, "Value-Driven Behavior Is a Business Imperative in the Global Economy," April 22, 1999, Penn State University.

34   *"After six or seven half-day meetings":* Author interview.

34   *Schacht recalls, "We spent":* Speech by Henry Schacht, "Value-Driven Behavior Is a Business Imperative in the Global Economy."

35   *One senior executive was very:* Author interview.

35   *As one explained, Lucent is a far better:* Author interview.

35   *"Everything is being rethought":* John Keller, "Unlikely Team: An AT&T Outsider and a Veteran Join to Run New Spin-off," *Wall Street Journal,* October 14, 1996.

CHAPTER FOUR

PAGE

39   *"Don't you mean 'Lucid'?":* Author interview.

40   *Even then, McGinn did not forget to credit:* Author interview.

40   *"When you bear one of the best known":* Lucent press release, "Systems and Technology Company Is Named Lucent Technologies," February 5, 1996.

42   *Yet neither was overly enthusiastic:* Author interview.

42   *"It's a horrible name":* Investors Business Daily, February 5, 1996.

43   *"I remember running downstairs to the hotel":* Schweig interview.

43   *Even Fitzgerald, who would be charged with putting:* Speech by Kathy Fitzgerald, "Building the Lucent Brand," *Fortune* seminar, 1997.

46   *Fiorina remembers that most people:* Patricia Sellers, "The 50 Most Powerful Women in American Business," *Fortune,* October 12, 1998.

47   *As one senior executive remembers:* Author interview.

47   *He told them, "If you come to Lucent":* Author interview.

49   *"Remember," he said, "this is the company":* Sellers.

49   *"As president of Lucent's consumer-products":* David Hamilton and Rebecca Blumenstein, "Hewlett-Packard Taps Outsider as President and Chief Executive," *Wall Street Journal,* July 20, 1999.

51   *She says, "I wanted desperately to make":* Carly Fiorina, "Making the Best of a Mess," *New York Times,* September 29, 1999.

51   *Putting systems in place was a morass:* Ibid.

52   *"Lucent is about what happens when you see the possibilities, not the limitations":* Nikki Goth Itoi, "Profile: Carly Fiorina, the Woman Behind Lucent's IPO," *Red Herring,* December 1998.

52   *"Carly is wickedly smart":* Sellers.

52   *"She had little experience with finance":* Ibid.

52   *Fitzgerald remembers Fiorina clearly:* George Anders, *Perfect Enough* (New York: Portfolio, 2003).

53   *"People thought we would sink like":* Itoi.

56   *"She [got] so psyched in these meetings":* Peter Burrows, *Backfire* (Hoboken, N.J.: John Wiley & Sons, 2003), p. 100.

56   *"That's how much she cared about":* Lusk interview.

57   *And as the china clinked and the glasses:* Schacht, IPO presentation, New York City.

57   *"Forget sexy services," she told them:* Fiorina, IPO presentation, New York City.

59   *The day ended with Fiorina and Lusk:* Burrows, p. 101.

59   *"We can't believe the attention":* John Keller, "Unlikely Team: An AT&T Outsider and a Veteran Join to Run New Spin-off," *Wall Street Journal,* October 14, 1996.

60   *McGinn once remarked:* Paul Kapustka, "Big Fish: McGinn Walks the Walk," *Red Herring,* June 29, 1999.

61   *"We set out to make our advertising unique":* Fiorina speech.

61   *First, he told Lucent employees:* Speech by Henry Schacht, "Value-Driven Behavior Is a Business Imperative in the Global Economy," April 22, 1999, Penn State University.

61   *"I probably worked just as hard at AT&T":* Jerry Useem, "The New Entrepreneurial Elite," *Inc.,* December 1997.

63   *Both men wandered the company:* Author interview.

63   *In 1997, Lucent sent questionnaires:* "Survey Shows Upbeat Pulse," *Lucent* magazine, March 1997.

66   *Bell Systems' first attempt to establish a research:* Robert Burderi, *Engines of Tomorrow* (New York: Simon & Schuster, 1996), p. 66.

67   *Robert Lucky, a Bell Labs:* Robert Lucky, Telcordia bio, www.telcordia .com/research/whoweare/welcome.html.

67   *"There was no specific charter":* Robert Buderi, "Lucent Ventures into the Future," *Technology Review,* November/December 2000.

68   *Bela Julesz, a Hungarian refugee who:* Jeremy Bernstein, *Three Degrees Below Zero: Bell Labs in the Information Age* (New York: Scribner, 1984).

68   *Arno Penzias, Nobel Prize recipient:* Ibid., p. 232.

69   *Penzias felt that radical changes were:* Buderi, p. 27.

69   *Brinkman remembers that:* Buderi, p. 37.

70   *He observed, "We could have gone anywhere":* Jay A. Conger, Gretchen M. Spreitzer, and Edward E. Lawler, *The Leader's Change Handbook* (San Francisco: Jossey-Bass, 1999), p. 18.

70   *In his earliest research he addressed:* Steve Levy, "Lucent Technologies: Patents for Profits," Salomon Brothers, November 1997.

71   *Levy labeled it the:* Levy interview.

71   *From his vantage point as close observer:* Ibid.

71   *"When Lucent came about it scared":* Paul Keegan and Katie Avoy, "Research at the Speed of Light: Lucent Technologies Recharges Bell Labs," *Upside,* February 1999.

71   *"A lot of people were looking":* Buderi, p. 270.

72   *As one former researcher noted:* Author interview.

72   *McGinn was already a convert:* Keegan and Avoy.

72   *For some, the answer was "no.":* Erik Schatzker, "Lucent's Fall from Grace," *Bloomberg News,* September 29, 2000.

73   *"In the past, research has come up":* Jason Krause, "Some Win, Some Lose in the Acquiring Game," TheStandard.com, August 7, 2000.

73   *"Some of these have moved at a speed that has sort of dazzled us":* Buderi, p. 270.

74   *Penzias noted the change in:* Peter Coy, "Commentary: The Lessons of the AT&T Break-up," *BusinessWeek,* November 22, 1999.

75   *Andy Grove, the legendary chairman:* Andy Grove, *Only the Paranoid Survive* (New York: Doubleday, 1996).

75   *Publicly, Lucent's unwavering belief:* Brian Taptich, "Lucent Unbound," *Red Herring,* August 1, 1998.

75   *"Every time something happens in the industry":* Brian Bergstein, "Bell Labs Struggles to Guard Its Legacy," *Toronto Star*/Associated Press, December 2, 2002.

CHAPTER FIVE

PAGE

77   *"A key to Lucent's success has been":* Lucent Press release, December 22, 1999.

77   *In his earliest report covering Lucent:* Steve Levy, "Lucent Technologies: Patents for Profits," Salomon Brothers, November 1997.

78   *"Business just exploded":* Andrew Kupfer, "Is Lucent Really as Good as It Seems?" *Fortune,* May 26, 1997.

78   *Envy had seeped into the relationship:* Author interview.

81   *Her seat companion overheard her:* Truax interview.

81   *He wrote, "The old switching business":* Dickson, internal memo to McGinn.

82   *McGinn proudly announced in 1998:* Brian Taptich, "Lucent Unbound," *Red Herring,* August 1, 1998.

82   *"We were taking half of AT&T's people":* Jay A. Conger, Gretchen M. Spreitzer, and Edward E. Lawler, *The Leader's Change Handbook* (San Francisco: Jossey-Bass, 1999), p. 24.

84   *Schacht was proud of protecting his employees:* Hedrick Smith, "Running with Bulls," PBS interview, www.hedricksmith.com/site_bottomline/html/schacht.html.

84   *McGinn was unabashedly ecstatic:* Taptich.

87   *As industry analysts Bart Stuck and Michael Weingarten noted:* Bart Stuck and Michael Weingarten, "The Unfashionable Service Provider," *Business Communications Review,* February 2001.

89   *Mel Cohen, who recently retired as Lucent's:* Paul Keegan and Katie Avoy,

"Research at the Speed of Light: Lucent Technologies Recharges Bell Labs," *Upside,* February 1999.

CHAPTER SIX

PAGE

93 *"Since 1995, we've taken a $20 billion business":* Nancy DuVergne Smith, "Communications Market Will Hit $900B by 2003, Lucent Chair Predicts," *TechTalk-M.I.T.,* May 17, 2000.

93 Worth *magazine ranked him:* Chris McDougal, "50 Best CEOs," *Worth,* May 1999.

93 Time *described him as:* "Cyber Elite 1998 Top 50," *Time* www.time.com/ time/digital/cyberelite/list.html.

94 Fortune *said, "Lucent resembled":* Kupfer.

94 *Steve Levy, the Lehman Brothers analyst:* Steven Levy, "Lucent Technologies: Patents for Profits," Salomon Brothers, November 1997.

94 *As one industry commentator noted:* Brian Taptich, "Lucent Unbound," *Red Herring,* August 1, 1998.

95 *Despite jibes from journalists:* Henry Goldblatt, "Go West," *Fortune,* November 22, 1999.

97 *Scooping up thirty-eight smaller companies:* Peter Howe, "Lucent Exec Janet Davidson: Let's Make It Work," *Boston Globe,* July 23, 2001.

99 *It was difficult to anticipate what Lucent would do:* Paul Keegan and Katie Avoy, "Research at the Speed of Light: Lucent Technologies Recharges Bell Labs," *Upside,* February 1999.

99 *Bill O'Shea, in his role as executive vice president of corporate strategy:* Jason Krause, "Blinded with Science," *The Industry Standard,* March 26, 2001.

99 *As McGinn explained, "There are a lot":* Keegan and Avoy.

100 *"We came here to pursue the American dream":* David Ignatius, "Business's Bold New Face," *Washington Post,* April 27, 1998.

100 *"Once you have some food":* Gallery of Achievement, interview with Jeong H. Kim, Jackson Hole, Wyoming, May 23, 1998.

100 *"If we had a communications network":* Bill Holland, "Jeong Kim: Setting the Pace," *Washington Business Journal,* September 16, 2002.

101 *"I said to myself I would do this thing for two years":* Ibid.

101 *"They offered me a job as President of Carrier":* Gallery of Achievement.

103 *Cisco CEO John Chambers pointed out that the competitive environment:* John A. Byrne and Ben Elgin, "Cisco Shopped Till It Nearly Dropped," *BusinessWeek,* January 21, 2002.

103   *Dan Plunkett, a Mercer Delta consultant who worked closely:* Author interview.

104   *In Pat Russo's words:* Peter K. Jacobs, "Pat Russo: Focused on the Future at Lucent Technologies," *Harvard Business School Bulletin online,* June 1999.

104   *"Quite frankly, we missed the last generation of":* Taptich.

104   *Kim was brutally honest when he pointed:* Gallery of Achievement.

106   *McGinn did not agree:* Neill Weinberg, "Wired and Restless," *Forbes,* February 7, 2000.

106   *"Twenty billion later, Lucent still needs":* Peter Elstrom, "Lucent's Ascent," *BusinessWeek,* February 8, 1999.

107   *John Dickson, head of Lucent's Microelectronics division, later gave:* Confidential memo from John Dickson to Rich McGinn, July 6, 2000.

107   *One business leader who had made his case to McGinn:* Author interview.

108   *"This is a true case of the customer is always right":* "Lucent Completes Merger with Ascend Communications," Lucent press release, June 24, 1999.

109   *"If Ascend and Cascade had cultural issues":* Herb Greenberg, "Why Two Analysts Have Made Negative Calls on Lucent," The Street.com, February 8, 1999.

109   *"Ascend only had one speed and they":* Levy interview.

109   *"Our culture had always been just to say yes":* Rebecca Blumenstein, "Switching Over: Lucent Bets a Packet on Bridging Worlds of Phone and Data Traffic," *Wall Street Journal,* June 10, 1999.

110   *One former Ascend manager explained:* Author interview.

110   *Marc Schweig, senior vice president of sales of Lucent, remembers:* Author interview.

110   *"Ascend could have been an injection of new DNA into Lucent":* Author interview.

111   *"One of the things they [acquired companies] bridled under":* Author interview.

111   *"The customer had called the meeting and was angry":* Author interview.

112   *"The Lucent management folks went on to discuss":* Author interview.

112   *Hendren first addressed his new Lucent employees wearing:* Author interview.

112   *"I believe in you more than you do in yourself":* Author interview.

113   *"Mike Hendren was really trying to help me":* Author interview.

113   *As one noted, "There has rarely been a collection of sales reps":* www.lightreading.com/boards/message.asp?msg_id=25960.

113   *But not everyone was impressed:* Boxer interview.

114   *One senior Lucent officer remembers:* Author interview.

115 *McGinn had no illusions about the essential ingredient:* Paul Kapustka, "Big Fish: McGinn Walks the Walk," *Red Herring,* June 29, 1999.

115 *"We all had opportunities outside of Lucent,":* Krause.

115 *At the time of the merger, analyst Levy observed:* Steven Levy, "A Tale of Two Deals," Lehman Brothers, January 20, 1999.

115 *They were the yin and yang:* Laurie Falconer, "Zhone Way or the Highway: How to Lead in Local Loop Infrastructure," *SiliconIran,* Issue 3.

116 *"I love to work. I might be":* Daniel S. Levine, "The Story with Morey: CEO of the Zhone Startup Tells All," *San Francisco Business Times,* March 27, 2000.

117 *Any expectation that Lucent could swallow Ascend:* Author interview.

117 *"The problem with most people from Lucent":* Author interview.

118 *When it became clear that nothing would come:* Author interview.

121 *"Why would someone want to create a fairly good":* Author interview.

122 *Lucent management grew fond of saying:* "Lucent Lessons for 2000," Bullet points for Henry Schacht, *Fortune* interview, January 4, 2001.

122 *Plenty of small companies are able to expand their businesses:* Russ deposition, May 10, 2002, p. 32.

123 *Steven Levy conceded that fantasy had:* Steven Levy, Sender Cohen, and Andrea Green, "Solomon Smith Barney Lucent Technologies Research Report," April 27, 1998.

123 *As he put it, "I wanted everyone":* Alice LaPlante, "The Man Who Built Lucent," *Electronic Business,* December 1998.

125 *This is an opportunity cost, but Harris Collingwood:* Harris Collingwood, "The Earnings Game: Everyone Plays, Nobody Wins," *Harvard Business Review,* June 2001.

126 *Arthur Levitt, while still chairman of the SEC:* Arthur Levitt, speech at the Economic Club of Washington, April 6, 2000.

126 *"Lucent is using several accounting practices which":* Smith Barney call notes, Tony Langham, February 26, 1997.

127 *"We believe it is important for investors to understand":* Ibid.

127 *The day after Langham disclosed his report:* John Keller, "Lucent Stock Drops on Analyst's Report, But Firm Defends Financial Practices," *Wall Street Journal,* February 26, 1997.

127 *Later, in conversations with the Hewlett-Packard board:* George Anders, *Perfect Enough* (New York: Penguin Books), p. 118.

128 *"With that kind of thinking, you'll be able to":* Jason Krause, "He Came from the North," *Industry Standard,* September 11, 2000.

130 *"We've talked about it very aggressively":* Ben Heskett, "Cisco, Lucent Won't Partner," Cnet_news.com, May 20, 1998.

130   *Later, McGinn's own managers and employees would:* Author interview.

131   *As the stakes rose with the stock price, the process:* McGinn deposition under oral examination, *Nina Aversano v. Lucent Technologies,* Civil Action No. MID-L-10004-00, Superior Court of New Jersey, p. 123.

132   *McGinn's message was this:* Author interview.

132   *McGinn's audience was ecstatic:* Alex Berenson, Simon Romero, and Seth Schiesel, "The Genesis of a Giant's Stumble," *New York Times,* January 21, 2001.

CHAPTER SEVEN

PAGE

133   *"Being driven by growth, growth, growth":* Erik Schatzker, "Lucent Investors Aren't Giving Up on CFO Hopkins Yet," Bloomberg News, February 21, 2001.

134   *Ben Verwaayen, former co–chief operating officer:* Peter Elstrom, "Lucent's Ascent," *BusinessWeek,* February 8, 1999.

135   *Orders that could not be booked, Levy posited:* Steven Levy, "Lucid Views on Lucent," Lehman Brothers, February 1999.

136   *"Four years ago, the businesses that became Lucent":* Chris McDougal, "50 Best CEOs," *Worth,* May 1999.

136   *Following in Schacht's footsteps was not an easy task:* "Richard A. McGinn: Light Speed," *Newsweek,* January 11, 1999.

137   *He remembers that "there was a period of time":* Meg McGinity and Dawn Bushaus, "The Selling Out of Innovation," *The Net Economy,* October 29, 2001.

137   *"It's like a jailbreak," explained Dan Smith:* Erik Schatzker, "Lucent's Fall from Grace," Bloomberg News, September 29, 2000.

138   *The turnover rate was 20 percent:* Julekha Dash, "Lucent Enhances Employee Perks," Itworld.com, October 30, 2000.

138   *"Customers . . . are increasingly dissatisfied with our performance":* Henry Schacht Opening Remarks, Lucent Technologies Officers' Meeting, Parsippany, New Jersey, November 8–9, 2000.

139   *He had spent his entire tenure at Lucent:* Alex Berenson, Simon Romero, and Seth Schiesel, "The Genesis of a Giant's Stumble," *New York Times,* January 21, 2001.

139   *"There were several softswitch efforts":* Ibid.

140   *One consultant who worked with management:* Peter Burrows, *Backfire* (Hoboken, N.J.: John Wiley & Sons, 2003), p. 108.

140   *"I remember [Fiorina's husband] Frank":* JoAnn S. Lublin and Rebecca

Blumenstein, "In This Family She's the CEO and He's at Home," *Wall Street Journal*, July 22, 1999.

140 *But Fiorina ignored them, following Schacht's advice:* George Anders, *Perfect Enough* (New York: Penguin Books, 2003), p. 53.

140 *She presented him with an option that:* Author interview.

140 *"Rich was a strategy person, often focused":* Author interview.

141 *He says, "Given Carly's record of achievement":* Plunkett e-mail.

142 *"It's about restlessness," said McGinn:* Neil Weinberg, "Wired and Restless," *Forbes*, February 7, 2000.

142 *"I am over the line":* Author interview.

143 *Schacht recalls the debate that took place over:* Ron Insana, "Can He Save Lucent?," *Money*, September 17, 2001.

144 *Joe Nacchio, erstwhile CEO of Qwest and senior AT&T:* Neil Weinberg and Scott Woolley, "Telecomeback," *Forbes*, January 21, 2002.

144 *Belief in the new economy, with its twin tenets:* Testimony of Chairman Alan Greenspan, State of the Economy, Before the Committee on Ways and Means, U.S. House of Representatives, January 20, 1999.

145 *Looking back from the vantage point of 2002:* McGinn deposition under oral examination, *Nina Aversano v. Lucent Technologies*, Civil Action No. MID-L-10004-00, Superior Court of New Jersey, p. 96.

146 *"Rich, there is a lot of confusion in the system,"* Memo from Ben Verwaayen to Rich McGinn, August 20, 1999.

146 *"The pervasive culture at Lucent was":* Author interview.

147 *Schacht remembers the problem as:* Henry Schacht Opening Remarks, p. 12.

147 *Lucent CFO Deborah Hopkins noted that such practices:* E-mail, Hopkins to Aversano, August 15, 2000, from *Nina Aversano v. Lucent Technologies*, Civil Action No. MID-L-10004-00, Superior Court of New Jersey.

147 *Nina Aversano, a senior sales executive at Lucent:* Nina Aversano, confidential deposition under oral examination in *Nina Aversano v. Lucent Technologies*, Civil Action No. MID-L-10004-00, Superior Court of New Jersey, p. 346.

148 *The situation deteriorated quickly:* Ram Charan and Jerry Useem, "Why Companies Fail," *Fortune*, May 27, 2002.

148 *Every quarter the pressure came down to the wire:* Lusk interview.

148 *As salespeople told one another:* Author interview.

148 *"Before I had left I told some of the people":* Author interview.

150 *Lucent's focus remained one-dimensional, as Microelectronics:* Dickson internal memo to McGinn, July 2, 2000.

152 *"We need serious teeth in inventory reduction programs":* Ibid.

152 *He told his boss, "We have trained":* Ibid.

154   *"I am old-fashioned about finance," he said:* Lusk interview.

154   *"This just didn't make business sense to me":* Ibid.

155   *One finance official who soon began to worry:* Author interview.

155   *But for salespeople it is a very dangerous drug:* Scott Moritz, "At Lucent, Staring into the Abyss," TheStreet.com, January 24, 2001.

156   *As one former Lucent salesperson noted:* Author interview.

156   *"It was high-yield heroin":* Peter Elstrom and Heather Timmons, "Telecom Meltdown," *BusinessWeek,* April 23, 2001.

157   *Phil Harvey, editor of an industry Web site:* "Vendor Financing," Lightreading .com, December 6, 2000.

157   *"The financing became the go/no go on the deal":* Author interview.

158   *"We think there's a natural boundary":* "Vendor Financing."

158   *CFO Hopkins said, "I think it's how we'll keep the velocity":* Scott Moritz, "Facing Investor Questions, Lucent Offers Few Answers," TheStreet.com, October 23, 2001.

158   *"If they want to be banks," analyst Levy noted:* Scott Moritz, "Cisco, Lucent and Nortel: Prime Lenders for the Network Buildout," TheStreet.com, November 8, 2000.

159   *"Do we need a 28th [digital subscriber line] network provider in Brooklyn?":* Ibid.

160   *What is clear is that each company:* Lucent press release, "Lucent Technologies Reports Results of Operational and Financial Review," December 21, 2000.

161   *It was affected by adverse:* Bart Stuck and Michael Weingarten, "Fixed Wireless—Do the Numbers Really Work?" *Business Communications Review,* June 2000.

162   *"Winstar is a focused, aggressive competitor":* Craig Bicknell, Wired News.com, October 22, 1998.

162   *One treasury official remembers, "People from treasury were":* Author interview.

163   *Its report took aim at a well-known short seller:* Jack Grubman, "WCII: Contrary to Other Opinions, We Believe WCII Has Ability to Finance Debt," *SalomonSmithBarney,* March 9, 2001.

164   *The mudslinging began in earnest:* Winstar Press release, April 18, 2001.

164   *Lucent argued that the responsibility:* Jim Wagner, Internet news.com, April 17, 2001.

165   *Thrilled with the possibilities, he wrote:* Robyn Roberts, *Lucent* magazine, November 1996.

167   *McGinn was confident about the course the business:* McGinn remembers: McGinn deposition under oral examination. *Nina Aversano v. Lucent Tech-*

nologies, Civil Action No. MID-L-10004-00, Superior Court of New Jersey, p. 53.

167   *At Lucent's analysts meeting in November 1999:* Ibid.

167   *No one at that meeting was shouting for OC-192:* Ibid.

167   *Warburg Dillon Reed analyst Nikos Theodosopoulos pointed out:* Nikos Theodosopoulos, "Warburg Dillon Reed Report on Lucent Technologies," November 12, 1999.

168   *"There was a risk that our equipment would get too hot":* Christine Y. Chen, "How Nortel Stole Optical," *Fortune,* October 2, 2000.

168   *"Joe said to me, 'I have all this fiber and I need to light it up' ":* Ibid.

168   *"We wanted a pipe that would carry 10 gigabits without":* Vicki Contavespi, "Pipe Dreams," *Forbes,* March 1, 1999.

169   *"Your enemy has a machine gun and you have a pistol":* Author interview.

169   *"They [Nortel] have gotten billions of dollars":* Deborah Solomon and Shawn Young, "Lucent Pays for Wrong Bet in Fiber-Optic Technology," *Wall Street Journal,* October 16, 2000.

169   *Reviewing Lucent's troubles with OC-192 from the vantage point:* Douglas Harbrecht, "We Tried to Run Faster Than We Were Capable of," *Business-Week,* November 6, 2000.

170   *Schacht proffered no excuses: "And shame on us":* Douglas Harbrecht, "Lucent's Henry Schacht: 'Shame on Us—I Mean Us, Big Us!' " *Business-Week,* October 26, 2000.

170   *One employee from Lucent's North Andover, Massachusetts:* Fifth Consolidated and Amended Class Action Complaint, in Re Lucent Technologies, Inc., p. 41.

171   *One nervous analyst asked her, "How much of a stretch":* Author interview.

171   *Nina Aversano and Pat Russo made an internal presentation:* "Service Provider Networks North America 2000 Operations Plan." Lucent documents.

172   *One senior manager within Lucent:* Author interview.

172   *Another senior manager remembers that:* Author interview.

172   *Internal documents from mid-November:* "Q1 2000," *Outlook Review,* November 19, 1999.

172   *Another top manager recalls meetings:* Author interview.

172   *A private memo from CFO Don Peterson to other top:* Interoffice Memo to senior Lucent Management from D. K. Peterson, date: November 3, 1999, Re: FY2000—First Quarter Gap Closure.

173   *"They offered discounts not only for that quarter but also on stuff":* Alex Berenson, Simon Romero, and Seth Schiesel, "The Genesis of a Giant's Stumble," *New York Times,* January 21, 2001.

173   *Schacht looked back upon these sales practices with misgivings:* Ibid.

174　*On January 2, 2000, McGinn told his board:* Private board of directors notes draft, January 2, 2000, 1 P.M.

174　*While no one had yet accused Lucent of illegality:* Stephanie N. Mehta and Elizabeth MacDonald, "Lucent Reports 51% Decline in Earnings," *Wall Street Journal,* January 21, 2000.

CHAPTER EIGHT

PAGE

179　*"Think of this as a new season":* Dennis K. Berman and Rebecca Blumenstein, "Major Business News Behind Lucent's Woes: Push to Meet High Revenue Goal," *Wall Street Journal,* March 29, 2001.

179　*"A quarter of a trillion dollars lost in value":* Author interview.

180　*"This is not a market issue," he insisted:* Kevin Petrie and Scott Moritz, "Lucent's Woes: 'Not a Market Issue,' " TheStreet.com, January 6, 2000.

180　*In a market where service providers were desperate:* Ibid.

181　*"Jealousy developed that people who only a few years":* Author interview.

181　*"Resentment was such that pulling back their business":* Author interview.

182　*"There's no fraud here," Levy said:* Joe McGarvey, Tom Steinert-Threlkeld, and Carol Wilson, "Lucent Numbers Not Adding Up," ZDNet.com, February 21, 2000.

182　*"They were trying to grow their business 20 percent":* Ibid.

182　*Even after listening to all of management's arguments:* Ibid.

183　*"He's almost physically ill if you bring him":* Ed Sussman, "50 Best CEOs," *Worth,* May 2000.

183　*Between 1989 and 1998 the number of individuals holding:* "The Investing Public," New York Stock Exchange.

185　*As he explained, "specifically the microprocessor":* Remarks by Alan Greenspan, Federal Reserve Board, April 11, 2000.

186　*On the final day of the century:* Thomas Petzinger, Jr., "So Long Supply and Demand. There Is a New Economy Out There and It Looks Nothing Like the Old One," *Wall Street Journal,* December 31, 1999.

188　*"It was okay with Nina if you took a risk":* Truax interview.

188　*Salespeople loved it. Truax remembers:* Ibid.

189　*Lucent was far from alone in this quest:* Speech by SEC Chairman: Arthur Levitt, "Quality Information: The Lifeblood of Our Markets," New York City, October 18, 1999.

191　*She needed a "personal pledge":* Author interview.

191　*"We went a little further than we should have":* Author interview.

192   *The SEC, in filing its case:* SEC civil complaint, May 17, 2004, p. 10; *SEC v. Lucent Technologies, Nina Aversano et al.*

192   *Another sales manager explained:* Author interview.

192   *"The discussion focused only on the current quarter":* Author interview.

192   *"My mother loves to party, loves people":* Patricia Sellers, "Behind Every Successful Woman There Is . . . A Woman. The mothers of the women on the Power 50 aren't the cookie-baking, stand-by-your-man type of mom. They are gutsy, creative iconoclasts. Like mother, like daughter," *Fortune,* October 25, 1999.

192   *With this position, Hopkins:* Patricia Sellers, "The 50 Most Powerful Women in American Business," *Fortune,* October 12, 1998.

193   *McGinn publicly stated that:* Ibid.

194   *She hammered at this point from:* Hopkins direct under oral examination in *Nina Aversano v. Lucent Technologies,* Civil Action No. MID-L-10004-00, Superior Court of New Jersey, p. 136.

194   *While Hopkins began with a burst of enthusiasm:* Erik Schatzker, "Lucent's Fall from Grace," Bloomberg News, September 29, 2000.

195   *"Our ultimate report card is the market":* Ibid.

195   *In April, she sent an e-mail to others:* E-mail from Aversano to Russo and others, April 4, 2000.

196   *Mercer Delta consultant David Nadler:* Nadler interview, June 14, 2003.

197   *"Part of what happened this quarter":* Erik Schatzker, "Lucent Investors Aren't Giving Up on CFO Hopkins Yet," Bloomberg News, February 21, 2001.

197   *Hopkins vowed, "It's not going to happen a third time":* Tom Johnson, "Struggling Lucent Replaces Exec," Newark *Star-Ledger,* May 7, 2001.

197   *"I don't have to tell you all how disappointing":* Transcript of Lucent Technologies conference call, January 26, 2000.

198   *"There is a good reason to be skeptical":* Scott Moritz, "Lucent Gets Lashed After Another Shortfall, and Heads May Roll," TheStreet.com, July 20, 2000.

199   *In September 2000, Craig Barrett, Intel's CEO:* Author interview.

200   *"The idea that we could be the leader in 18 different":* Erik Schatzker, "Lucent's Fall from Grace," Bloomberg News, September 29, 2000.

201   *She said, "We need to think of the issue":* E-mail from Russo to Aversano and others, July 25, 2000.

202   *"We were running on systems and processes":* "Lucent Lessons for 2000," Henry Schacht Bullet Points for *Fortune* article, January 4, 2001.

202   *"Information doesn't really flow through them":* Michael Copeland, "Lucent awakes from optical delusions," *Red Herring,* December 22, 2000.

202   *By "pulling forward" sales:* SEC civil complaint, May 17, 2004, p. 2; *SEC v. Lucent Technologies, Nina Aversano et al.*

202   *Deborah Harris, a vice president in Lucent's:* Fifth Consolidated and Amended Class Action Complaint, Civil Action number 00-621.

203   *When asked whether she had been aware:* Hopkins direct under oral examination in *Nina Aversano v. Lucent Technologies,* p. 244.

203   *While he apologized in advance for anything:* John Dickson, "Confidential Brain Dump," July 6, 2000.

203   *Dickson's "brain dump," which runs:* Ibid.

204   *Lucent was fast losing ground, O'Shea explained:* Internal e-mail, William O'Shea to Rich McGinn, Subject: Officers Meeting Thoughts, August 14, 2000.

206   *Kim was blunt with his assessment:* Neil Weinberg, "Wired and Restless," *Forbes,* February 7, 2000.

206   *Later, when reviewing the range:* Alex Berenson, Simon Romero, Seth Schiesel, "The Genesis of a Giant's Stumble," *The New York Times,* January 21, 2001.

207   *Although McGinn continued to support:* McGinn deposition under oral examination in *Nina Aversano v. Lucent Technologies,* Civil Action No. MID-L-10004-00, Superior Court of New Jersey, p. 159.

207   *In giving her such increased responsibilities:* Ibid.

207   *Others on McGinn's team had similar sentiments:* Fitzgerald deposition under oral examination in *Nina Aversano v. Lucent Technologies,* Civil Action No. MID-L-10004-00, Superior Court of New Jersey, p. 217.

208   *Dickson also suggested that McGinn needed to replace:* Dickson, "Confidential Brain Dump," July 6, 2000.

208   *McGinn remembers . . . "collectively—and individually":* McGinn deposition, *Nina Aversano v. Lucent Technologies,* Civil Action No. MID-L-10004-00, Superior Court of New Jersey, p. 259.

209   *McGinn believed that Russo was:* McGinn deposition under oral examination in *Nina Aversano v. Lucent Technologies,* Civil Action No. MID-L-10004-00, Superior Court of New Jersey, p. 175.

209   *For McGinn this was an enormous frustration:* Henry Schacht deposition under oral examination in *Nina Aversano v. Lucent Technologies,* Civil Action No. MID-L-10004-00, Superior Court of New Jersey, p. 29.

209   *"Then if that's the case, there's really":* Pat Russo deposition under oral examination in *Nina Aversano v. Lucent Technologies,* Civil Action No. MID-L-10004-00, Superior Court of New Jersey, p. 19.

209   *McGinn's leadership was also in question:* Pat Russo deposition under oral examination in *Nina Aversano v. Lucent Technologies,* Civil Action No. MID-L-10004-00, Superior Court of New Jersey, pp. 32, 40.

210   *Aversano remembers from her conversation with Russo:* Nina Aversano, confidential deposition under oral examination in *Nina Aversano v. Lucent Technologies,* Civil Action No. MID-L-10004-00, Superior Court of New Jersey, p. 317.

210   *"It's not that he was defeated":* Author interview.

211   *He needed their renewed commitment:* Hopkins direct under oral examination in *Nina Aversano v. Lucent Technologies,* Civil Action No. MID-L-10004-00, Superior Court of New Jersey, p. 132.

211   *"The feeling was that Rich had become":* Author interview.

211   *"Why should I believe you?":* Author interview.

211   *"This is absolutely the number":* Author interview.

211   *Hopkins later explained:* Hopkins direct under oral examination in *Nina Aversano v. Lucent Technologies,* Civil Action No. MID-L-10004-00, Superior Court of New Jersey, p. 243.

211   *But the problem in Hopkins's view was even larger:* Ibid.

212   *Aversano recalls an entirely different situation:* Nina Aversano, confidential deposition under oral examination in *Nina Aversano v. Lucent Technologies,* Civil Action No. MID-L-10004-00, Superior Court of New Jersey, p. 342.

212   *What Aversano failed to say:* SEC civil complaint, May 17, 2004, p. 3; *SEC v. Lucent Technologies, Nina Aversano et al.*

212   *McGinn disagrees and has said:* Dennis K. Berman and Rebecca Blumenstein, "Behind Lucent's Woes: Push to Meet High Revenue Goals," *Wall Street Journal,* March 24, 2001.

213   *She remembers that on that call in August:* Nina Aversano, confidential deposition under oral examination in *Nina Aversano v. Lucent Technologies,* Civil Action No. MID-L-10004-00, Superior Court of New Jersey, p. 385.

213   *Only three days earlier she had sent an e-mail:* E-mail from Nina Aversano to management group, Subject: 4Q Product Shortages, August 11, 2000.

213   *A member of her team, Marc Schweig:* E-mail from Marc Schweig to Jay Carter, August 14, 2000.

213   *When a salesperson came onto a call:* Author interview.

214   *According to McGinn, when he:* McGinn deposition under oral examination in *Nina Aversano v. Lucent Technologies,* Civil Action No. MID-L-10004-00, Superior Court of New Jersey, p. 299.

214   *She claims that she made herself:* Nina Aversano, confidential deposition under oral examination in *Nina Aversano v. Lucent Technologies,* Civil Action No. MID-L-10004-00, Superior Court of New Jersey, p. 375.

215   *In McGinn's sworn deposition:* McGinn deposition under oral examination in *Nina Aversano v. Lucent Technologies,* Civil Action No. MID-L-10004-00, Superior Court of New Jersey, p. 294.

215   *As a result, and until they had further:* Letter from Richard Rawson to Nina
       Aversano, November 28, 2000.

215   *Aversano struck side agreements:* SEC civil complaint, May 17, 2004,
       p. 11; *SEC v. Lucent Technologies, Nina Aversano et al.*

215   *Lucent's management felt sorely betrayed by Aversano:* Author interviews.

216   *Aversano had been threatening to leave for months:* Author interviews.

217   *Moreover, they say, she had to:* Author interviews.

217   *"Her saying that she discovered":* Berman and Blumenstein, "Behind Lu-
       cent's Woes: Push to Meet High Revenue Goals," *Wall Street Journal,*
       March 29, 2001.

217   *A Lucent internal document dated September 21:* Attachment to e-mail from
       Richard Lanahan to Deborah Hopkins, James Lusk, Mark White, Catherine
       Carroll, Stephen Brockman, and Michael Rec, Subject: 2001 plan review,
       date September 21, 2001.

218   *Even though all of the market's goodwill:* Lucent press release, July 20,
       2000.

219   *"We have heard the comments made by":* Ben Heskett and Wylie Wong,
       "Lucent Support of Start-ups Could Point to Problems," Cnet.com, October
       11, 2000.

219   *McGinn remembers that, during Aversano's presentation:* McGinn deposi-
       tion under oral examination in *Nina Aversano v. Lucent Technologies,* Civil
       Action No. MID-L-10004-00, Superior Court of New Jersey, p. 269.

220   *The board was widely criticized for being:* Matthew Boyle, "The Dirty Half-
       Dozen: America's Worst Boards," *Fortune,* April 30, 2001.

221   *One senior executive says, "I remember on":* Author interview.

221   *Among his inner circle he had been:* Rawson, direct questioning deposition
       in *Nina Aversano v. Lucent Technologies,* Civil Action No. MID-L-10004-
       00, Superior Court of New Jersey, p. 163.

222   *One senior manager said, "I remember being on a boat":* Author interview.

223   *For Schacht, it was the end of a relationship he had:* Douglas Harbrecht,
       "Lucent's Henry Schacht: 'Shame on Us—I Mean Us, Big Us!,' " *Business-
       Week,* October 26, 2000.

224   *Controller Jim Lusk remembers speaking to him shortly:* Lusk interview.

CHAPTER NINE

PAGE

225   *"Our foundations are sound as our restructuring":* Simon Romero, "Lucent
       Credit Rating Falls," *New York Times,* February 13, 2001.

225   *"In spite of considerable effort and success in several":* Internal e-mail,

William O'Shea to Rich McGinn, Subject: Officers Meeting Thoughts, August 14, 2000.

226   *"We developed a reputation for calling":* Erik Schatzker, "Lucent Shareholders Grill CEO Schacht on Stock, Board," Bloomberg News, February 21, 2001.

226   *He felt that the board had acted correctly but admitted:* Alex Berenson, Simon Romero, and Seth Schiesel, "The Genesis of a Giant's Stumble," *New York Times,* January 21, 2001.

226   *In hindsight Schacht felt that McGinn had lost:* Schacht deposition under oral examination in *Nina Aversano v. Lucent Technologies,* Civil Action No. MID-L-10004-00, Superior Court of New Jersey, p. 65, November 5, 2001.

226   *"Stock price is a byproduct":* Stephanie Mehta, "Lessons from the Lucent Debacle," *Fortune,* February 5, 2001.

226   *"I think we tried to run faster than we were capable":* Douglas Harbrecht, "We Tried to Run Faster Than We Were Capable of," *BusinessWeek,* November 6, 2000.

227   *After a brief analysis Schacht determined that Lucent's problems:* Ibid.

227   *"The parking lot fills at 9 o'clock":* Scott Moritz, "Resistible Force, Meet Immovable Object: Lucent Board Has CEO Issue," The Street.com, October 11, 2000.

227   *"It all boils down to kind of one fundamental thing":* Seth Schiesel, "How Lucent Stumbled: Research Surpasses Marketing," *New York Times,* October 16, 2000.

228   *He told the scientific team:* Author interview.

229   *After that, "We're going to talk about values":* Henry Schacht Opening Remarks, Lucent Technologies Officers' Meeting, Parsippany, New Jersey, November 8–9, 2000.

229   *"Just so we all understand":* Ibid.

230   *Going backward could not be the answer:* Author interview.

230   *And very much in keeping with his philosophy:* Henry Schacht Opening Remarks, Lucent Technologies Officers' Meeting, November 8–9, 2000.

230   *"What has happened to us is that our execution and processes":* Ibid.

230   *In two weeks of discussions with colleagues:* Ibid.

230   *The SEC would later charge:* SEC civil complaint, May 17, 2004, p. 27; *SEC v. Lucent Technologies, Nina Aversano et al.*

231   *He publicly exposed a list of sins:* Ibid.

231   *Major customers, Schacht relayed, had said:* Ibid.

231   *The most painful revelation had come from a client:* Ibid.

232   *As one executive remembers, "It was an environment":* Author interview.

232   *A postscript to this anecdote:* SEC civil complaint, May 17, 2004, p. 21 and 33; *SEC v. Lucent Technologies, Nina Aversano et al.*

232   *She told him emphatically:* Henry Schacht Opening Remarks.

233   *After Schacht's long-winded twenty-two-page speech, he broke:* Ibid.

233   *In the months after Schacht's return, a senior colleague:* Author interview.

234   *Winstar's management was not pleased and told:* Author interview.

234   *The Lucent executive involved:* E-mail from Carole Spurrier to Rich McGinn, Deborah Hopkins, and Nina Aversano, Subject: BellSouth billing, September 21, 2000.

235   *Schacht promised the market whiter than white:* Ron Insana, "Can He Save Lucent?," *Money,* September 17, 2001.

235   *He was inundated by a barrage of questions:* Ibid.

235   *One senior regulator commented to* Fortune: Carol Loomis, "The Whistle-blower and the CEO," *Fortune,* June 23, 2003.

235   *Hopkins was keenly aware of this risk:* E-mail from Deborah Hopkins to Henry Schacht and Kathy Fitzgerald, Subject: coverage on last week, November 26, 2000.

236   *Schacht reflected, "In striving to sustain growth targets":* Berman and Blumenstein, *Wall Street Journal,* March 29, 2001.

237   *With optimism that would not be rewarded:* Erik Schatzker, "Lucent Shareholders Grill CEO Schacht on Stock, Board," Bloomberg News, February 21, 2001.

240   *"It's bad and there is no cavalry coming":* John Schwartz, "Another 7,000 Jobs to Be Cut as Lucent Reports More Losses," *New York Times,* July 24, 2002.

241   *The behavior followed a predictable pattern:* Charles P. Kindleberger, *Manias, Panics and Crashes: A History of Financial Crises* (New York: Wiley, 2000), p. 16.

243   *His pitch was one part nostalgia, one part patriotism:* CNN.com Europe, February 20, 2001.

244   *"We don't want to do this":* Mark Lake, Bloomberg News, February 17, 2001.

244   *Only those institutions that could find their:* Ibid.

244   *"This is the most extreme example we've seen so far":* Suzanne McGee, "Lucent Rewarded Leaders with Underwriter Roles," *Wall Street Journal,* February 28, 2001.

248   *She had not decoded the company's unwritten:* Author interview.

248   *Hopkins described herself as a:* Patricia Sellers, "The 50 Most Powerful Women in American Business," *Fortune,* October 12, 1998.

249   *She had had no idea of the depth of Lucent's problems before:* Dennis

Berman, "Lucent Finance Chief Hopkins Steps Down After One Year," *Wall Street Journal,* May 7, 2001.

249   *After Hopkins's departure the press was scathing:* Jay Chrepta, "A Short-Term Look at the Long-Term Business," *HBS Working Knowledge,* February 5, 2001.

251   *Schacht had displayed enthusiasm:* Nikhil Deogun and Dennis Berman, "Lucent, Alcatel Terminate Merger Talks," *Wall Street Journal,* May 30, 2001.

253   *"Why would we walk away from a deal":* Dennis Berman, *Wall Street Journal,* June 1, 2001.

253   *He says now, "Alcatel would have":* Author interview.

CHAPTER TEN

PAGE

255   *"This was once a great company, and it was something to be proud of":* Peter Howe, "Lucent Exec Janet Davidson: Let's Make It Work," *The Boston Globe,* July 23, 2001.

255   *PR head Kathy Fitzgerald reflects:* Author interview.

256   *Dan Plunkett, Lucent's longtime consultant:* Plunkett interview.

256   *"Without fresh blood at the senior management":* Scott Moritz, "Lucent's Vision Further Impaired," TheStreet.com, January 7, 2002.

257   *In his first week back as CEO he made it clear:* Douglas Harbrecht, "Lucent's Henry Schacht: 'Shame on Us—I Mean Us, Big Us!,' " *BusinessWeek,* October 26, 2000.

257   *"You walk away knowing that she":* Olgel Kharif, "Pat Russo: Lucent's Best Hope," *BusinessWeek,* May 27, 2003.

257   *"Since she has been there":* Ibid.

257   *Taking a line from Winston Churchill:* Ibid.

258   *Later, Lucent faced a fine:* Press release, "Lucent Settles SEC Enforcement Action Charging the Company with $1.1 Billion Accounting Fraud," SEC, May 17, 2004.

258   *In her lawsuit against Lucent, Aversano's lawyers had grilled Schacht:* Schacht direct under oral examination in *Nina Aversano v. Lucent Technologies,* Civil Action No. MID-L-10004-00, Superior Court of New Jersey, p. 60.

259   *As Russo looked back on her first:* "Pat Russo, Lucent Technologies," as told to Jason Meyers, *Telephony Online,* June 2, 2003.

260   *"It was a facility that was all-encompassing":* Andy Murray, "A Workplace for Generations," *Eagle-Tribune* (Merrimack Valley), February 23, 2003.

261   *"In the early 1980s, you needed 500 people to make copper":* David Wallace, "Out of Order," *Commonwealth Magazine,* Winter 2002.

261   *"Merrimack Valley is the premier optical networking manufacturing":* Lucent press release, June 7, 2000.

261   *"People would kill to get a job at Lucent":* Author interview.

261   *He recalled, "Our plant manager told us, in captive meetings":* Gary Nilsson, "Lucent Workers Fight Plant Closing to the Bitter End," *Labor Notes,* January 2002.

262   *"We thought that after the other plants closed":* Maddie Carrier interview.

262   *"Our plant manager held another captive meeting":* Nilsson.

263   *"We thought that was behind us, that was the end":* Maddie Carrier interview.

263   *"When the layoffs first started, Lucent":* Nilsson.

264   *"They lied, they lied, they lied":* Author interview.

264   *"This is really tough because most of these":* Molly Manchenton, "950 More Layoffs Jolt Lucent," *Eagle-Tribune* (Merrimack Valley), September 2001.

264   *"It feels funny to leave. I came in this morning":* Ethan Forman, "Saying Goodbye to Lucent," *Eagle-Tribune* (Merrimack Valley), June 2, 2001.

264   *Yet Russo was very careful to temper:* "Lucent Technologies Today," October 29, 2003.

INDEX

Access Management, 51
accounting fraud, 241–42
accounting manipulation, 125–32,
      174–77
Agere, 37, 198, 199, 200, 238, 245–47
Alcatel, 16, 81–82, 219, 236, 245
   proposed Lucent merger with, 248,
      249–53
Alcoa, 29, 38
Allaire, Paul, 38
Allen, Robert, 2, 9, 11, 13, 14, 15–19,
      27–28, 30, 31, 48, 59, 84, 199
Amazon, 59
Amber Networks, 115
American Express, 63–64
Ameritech, 12
AOL, 80
Ascend Communications, 105–18, 139,
      252
asynchronous transfer mode (ATM),
      equipment, 99–104, 105–18, 184
AT&T, 31–32, 78, 98, 101, 112–13,
      129–30, 138, 139, 141–42, 154,
      169, 187, 194, 247, 262, 264
   Access services division of, 51
   antitrust suits and, 12

Baby Bells' competition with, 14
business mistakes and failures of,
      16–18
competition and, 9, 12, 13, 14, 84–85
Computer Systems division of, 13
consent decree of 1956 and, 12
consent decree of 1982 and, 12
culture of, 21–38, 53–54, 60–63, 89,
      107
data networking and, 104
Data System Group of, 22
directors of, 28–29, 84
diversification attempts of, 13
divestiture of, 68
employees of, 9, 11, 61, 209, 249,
      260
executive compensation at, 84
Federal Communications
      Commission and, 10
global business of, 81
government break-up of, 9–19,
      84–85
government investigations of, 10–13
Information Systems division of, 68
information systems of, 201–2
International division of, 22

AT&T (*cont.*)
  Interstate Commerce Commission
    and, 10
  long distance business losses of, 13
  Long Lines business of, 70
  as Lucent customer, 174, 181, 196,
    220, 232, 236
  Lucent spin-off by, 5–19, 26–28,
    44–48, 50–59, 107, 137, 198, 245
  management talent in, 53
  MCI's challenge to, 12
  as monopoly, 5–6, 9–13, 27, 66,
    84–85
  Network Systems division of, *see*
    Network Systems (AT&T)
  opportunities missed by, 13
  optical technology abandoned by,
    166
  packet switching and, 89
  pricing power of, 10
  product cycles of, 62
  purchase of TCG by, 85–86
  regional carrier acquisitions by, 10
  shareholders of, 9, 13
  slowness to market of, 67, 89, 94–95,
    101
  stock price of, 18
  Sun Microsystems purchased by, 22
  vertical integration in, 9, 14–15
  see also Bell Labs; Bell Systems;
    Network Systems (AT&T)
AT&T Labs, 69
Atkins, Betsy, 222
ATT Wireless Services, 232
Avaya, 37, 81, 194, 198, 200, 209, 238
Aversano, Nina:
  CLEC market and, 188
  compensation plan and, 188
  conference calls led by, 189–92
  directors and, 210
  executive management's lack of
    confidence in, 208, 211
  Fiorina and, 188–89
  forecasting problems and, 192,
    211–12, 217
  fraudulent conduct under, 189
  as head of North American Service
    Provider Networks, 187, 207
  Hopkins and, 208, 211
  lagging sales of, 195–96, 213
  lawsuit against Lucent of, 210
  legacy product sales and, 171–72
  Lucent settlement with, 257–58
  McGinn's integrity questioned by,
    210
  McGinn's lack of confidence in,
    207–19
  misrepresentations by, 192
  paid administrative leave for, 215
  performance of, 208
  product shortages and, 213
  resignation submitted by, 214
  retirement of, 233
  sales management style of, 188, 191,
    213–14
  sales practices of, 147, 215
  Schacht's refusal to settle with,
    216
  SEC lawsuit and, 147, 189, 192, 215
  stock analysts and, 135, 187–88,
    214–15
  whistle blowing by, 215
  Winstar and, 202

Baby Bells, 7, 12, 14–15, 48, 53, 79,
    104, 107–8, 131–32, 137, 139,
    151, 240
  as aggressive competitors, 227
  CLECs and, 85–91, 227
  spending cuts by, 236

technological innovation and,
169–70, 184
telecom crash and, 253–54
bandwidth, 165–66, 168, 186
Bankruptcy Code, U.S., 164
Barden, David, 5
Barrett, Craig, 199–200
Barron, Robert, 120–21, 141–42
Bay Networks, 48, 105, 128
Bell, Alexander Graham, 10, 41, 66
optical technology and, 165
Bell Atlantic, 12
Bellcore, 68
Bell Labs (AT&T), 9, 14, 57, 88, 139,
199–200
break-up of, 68–69
culture of, 67–69
employees of, 68
financing of, 66–67
government regulation of, 10–12
history of, 65–75
insularity of, 67
intellectual chemistry at, 67–68
laser invented by, 165
management of, 68–69
optical technology and, 165
packet switching telephony and, 89
patents and innovations of, 10–11,
54, 56, 65–67, 80
power of, 67
product design of, 67
researchers' exodus from, 69
standards and quality at, 71–72
technological change and, 67
Bell Labs (Lucent), 112, 129, 198, 216
brain drain from, 137
Computing Sciences Research
Center of, 74
data networking and, 104, 107
downsizing of, 254

economic research of, 136, 143
innovation at, 73–74, 94, 104, 206
Merrimack Valley Works plant and,
261
optical technology and, 89, 165–71,
206
Packetstar IP Switch and, 107
R&D investment in, 70, 71–73,
74–75
remaking of, 69–75
standards and quality at, 71–72
BellSouth, 12, 196, 234–35
Bell Systems, 10, 66
Betamax, 165
Big Bang, 65–66, 69
biological computers, 72
Boeing, 193
Bosco, Harry, 205
Boston College, 194
Boxer, Lance, 25–26, 113, 139
Briere, Daniel, 42
Brinkman, William, 69, 137, 164–67
British Telecom, 37, 249, 255
Brown, Charlie, 12
BUFKANS, 138, 146
business cycles, 185–86
business investment, 185
Business Week, 99–100, 136
Buzzword Bingo, 187–92

C (computer language), 11
cable operators, 84–85
Cablevision, 196
Carr, Harry, 102, 257
Carrier, Maddie, 262–63
Carter, Jay, 212, 232
Cascade, 105–6
CBS, 29
cellular telephony, 65, 73–74
growth in, 79, 84

Center for Financial Research and
    Analysis, 174–75, 182
Chambers, John, 103, 106, 129–30, 204
Chase Manhattan, 29
Christensen, Clayton, 169–70
Chromatis, 118–21, 141–42
Churchill, Winston, 257
Ciena, 11, 53–54, 170
circuit switching, 67, 136
Cisco, 94
    acquisitions and, 75, 97, 103, 120
    Baby Bells and, 139
    Cerent acquisition by, 120
    as data networking market leader,
        105, 106, 200
    employee turnover at, 137–38
    investment quality of, 59
    Juniper Networks threat to, 106
    Lucent acquisition threat to, 119
    as Lucent competitor, 33, 60, 139,
        200
    Lucent's prospective partnership
        with, 129–30
    as model for Lucent, 27, 75
    optical system sales of, 201
    P/E of, 142
    research by acquisition model of,
        75
    vendor financing and, 158
Clarkson, Ken, 74
CLECs (Competitive Local Exchange
    Carriers), 62, 85–91, 103, 131–32,
    228
    Aversano and, 187–92
    cash needs of, 90, 142, 159–60
    collapse of, 236, 253–54
    credit quality of, 90
    vendor financing of, 90, 156–64
Cohen, Mel, 73, 89
Collingwood, Harris, 125

Collins Radio, 17
Commerce Department, U.S., 184
Communications Workers of America,
    261–63
Compaq, 163
CompuServe, 80
Condit, Phil, 193
Continental Airlines, 17
Corning, 245
Crawford, Curtis, 47
Credit Suisse First Boston, 163
Crisis in the Valley Committee, 263
Cummins Engine, 2, 28, 29, 30, 31, 32,
    33, 55, 228, 232

D'Amelio, Frank, 249
dark matter, 72
data networking, 21, 103–18, 197,
    200
Davidson, James, 255
Davidson, Janet, 97
days sales outstanding (DSO),
    150–51
deButts, John, 12
Dell Computer, 23
dense wavelength division multiplexing
    (DWDM), 53–54, 165–67
Dickson, John, 81, 107, 150, 151,
    203–4
"Dirty Half-Dozen" (Fortune), 220
"Doug Fluties," 194, 214
Drew, John, 141–42

EBITDA (earnings before interest,
    taxes, and depreciation
    allowance), 163
economy, global, 78–79
Ejabat, Mory, 109, 112, 115–16
Enron, 122, 241
Ericsson, 15–16, 126, 131–32

Europe, 16–17
Excel, 139
executive compensation, 54, 57–58, 84, 94, 122–23

Federal Accounting Standards Board, 97–98
Federal Communications Commission, U.S., 10, 85
Federal Reserve Board, 123, 143–45, 183–87
    business investment and, 185
    monetary policy and, 184
    productivity growth and, 185
fiber-optic networks, 15, 86–91, 164–71
    overcapacity of, 145–46, 186
    see also optical fiber; optical transport
Fiorina, Carly, 15–16, 35–36, 41–42, 49, 127, 138, 139–41, 173, 193–94
    Aversano and, 188–89, 191
    background of, 51–52
    Hewlett-Packard's hiring of, 140–41
    Lucent's advertising and, 60–61
    Lucent's IPO and, 52–59, 103
    MCI deal and, 139
    Winstar deal and, 162
    Yurie Systems acquisition and, 102
Fitzgerald, Kathy, 1–4, 17, 39–40, 42, 140, 179, 208, 225, 251, 255–56
5ESS (electronic switching system), 80–81, 95
Ford Foundation, 29, 38, 221
Fortune, 93–94, 101, 123, 140, 192–93, 202, 220, 235
401k plans, 109–10
fraud, accounting, 241–42

fund managers, 197, 243
Furukawa Electric, Lucent fiber business purchased by, 253

GAAP, 175–76
General Electric, 250
Gidron, Rafi, 118–21
Global Crossing, 158, 233
global teledensity, 81
Godfather III, The, 225
Greenspan, Alan, 123, 143–45, 184
    power of technology and, 185–86
Greybar, 215
Grinnell College, 22
Grove, Andy, 75
GTE, 81
Gulf War, 100

Hallmark, 63–64
Harris, Deborah, 202
Harvard Business Review, 125, 244
Harvard Business School, 13
Harvard College, 66
Harvey, Phil, 157
Hayes, Hammond, 66
Hayes, Samuel, 244
Hedren, Mike, 110, 112–13
Heidrick & Struggles, 220
Hewlett-Packard, 11, 27, 36, 37, 38, 63–64, 127
high-definition television, 65
Hills, Carla, 37–38
Holder, Robert, 199–200, 208
Holland, Royce, 156
Holzmann, Gerald, 72
Home Depot, 122
Hopkins, Deborah:
    Agere IPO and, 244
    as aggressive turnaround expert, 192
    ambitions of, 193

Hopkins, Deborah (*cont.*)
  audit committee confrontation with, 222
  Aversano's replacement and, 208
  bad debt and, 197
  bankruptcy rumors denied by, 247
  budgeting process overhaul by, 193
  candidness of, 247–48
  CLECs and, 197
  comments about McGinn and Aversano by, 211
  Condit's praise for, 193
  confrontational style of, 193
  credibility of, 197
  directors' complaints about, 247
  financial discipline of, 193
  friction with AT&T staff of, 194, 247
  growth-driven operations philosophy of, 133
  hired as CFO, 192–93
  illegal sales behavior and, 232
  Lucentspeak aversion of, 194
  McGinn's common vision with, 248
  optimistic financial projections of, 218–19
  pessimistic financial forecast by, 222
  "pull-ups" and, 203
  refinancing negotiated by, 243–49
  sales practices and, 147, 203, 232
  Schacht's disagreements with, 248
  surprises faced by, 193, 194, 197, 249
  vendor financing and, 158, 197

IBM, 13, 94, 165, 247
Illinois Bell, 22
inflation, 144–45
information technology, 185–86
initial public offerings (IPOs), 5–6, 40, 50–59, 64, 100, 102, 106, 107, 128–29, 130, 135, 171, 199, 245–47
  CLECs and, 86–91
Innovator's Dilemma (Christensen), 169–70
integrated circuits, 50
Intel, 75, 165, 199–200
International Network Services, 141–42
Internet, 124, 136, 183–84
Internet routers, routing, 104, 106–8, 118–19
Interstate Commerce Commission, U.S., 10
investment bankers, banking, 127–32, 135–36, 244
IP (internet protocol), 106–8, 184
"irrational exuberance," 184

JDS Uniphase, 59, 245
Jewel Co., 38
Johns Hopkins University, 100
J.P. Morgan Chase & Co., 158
Julesz, Bela, 68
Juniper Networks, 106–8, 158, 216
Justice Department, U.S., 10–12

Kanan, Joseph, 261
Kerrey, Bob, 61
Keynes, John Maynard, 159, 164
Kim, Jeong, 99–104, 141–42, 205–7
Kindleberger, Charles, 241
Knox, Wayne, 73
Kodak, 256
Kriens, Scott, 106–8, 158

Lacouture, Paul, 257
Lambda router, 73, 240, 261
Landor Associates, 40–44
Langham, Tony, 126–27

Laroia, Rajiv, 67–68
lasers, 65, 165–67
Lehman Brothers, 70–71, 93–94, 134–36
Levitt, Arthur, 126, 189
Levy, Steven, 70–71, 77–78, 93–94, 109, 123, 132, 134–36, 158, 159, 174, 181–82, 197, 256
Lewis, Drew, 37–38
Lightreading.com, 157
local vs. long distance calling, 12, 16–17
Loomis, Carol, 123
Lucent:
  Access Service division of, 172
  accountability and, 204, 224
  accounting practices of, 125–32, 134–35, 147–77, 181–82, 202–3, 212, 230–31, 242
  acquisitions of, 97–121, 133, 252
  advertising of, 6, 43–44, 56, 60–61
  Agere IPO of, 199, 245–47
  Ascend Communications purchase by, 105–18, 133, 252
  asset sales of, 253
  AT&T spin-off of, 5–19, 26–28, 44–48, 97, 107, 137
  ATM business of, 99–104, 105–18
  backlogs and, 153
  bad debt of, 150–51, 159, 197, 217–18
  bankruptcy rumors about, 247
  Bell Labs' importance to, 69–70
  BellSouth software deal with, 234–35
  bond ratings of, 245, 253, 254
  brain drain from, 121, 128–29, 137–41, 195, 227
  broadband business of, 198
  BUFKANS group in, 138–39, 146
  Business Communications Systems (BCS) division of, 48, 137, 169, 173, 194
  businesses shut down by, 238–40, 260–64
  Carrier Networks division of, 101
  cash crisis of, 242–45, 253
  CEO search of, 255–57
  change and, 77–91
  circuit-switching business of, 136, 151, 196–97, 218
  Cisco's partnership with, 129–30
  CLECs and, 136, 151, 157–64, 197, 217–18, 227–28, 240
  competition and, 37, 52–53, 81–82, 94–95, 97, 119, 136, 155, 156, 214
  Consumer Products group of, 48–49
  core business of, 9–10, 79–80
  cost cutting by, 172–73, 237–40, 254
  cost structure of, 82–84, 138
  credit rating of, 6, 239, 243, 253, 254
  crisis of, 202–54
  culture of, 7, 21–38, 45, 50, 53–54, 57–58, 60–65, 77, 90–91, 94, 109–18, 146–49
  customer service of, 103, 107–8, 161, 231
  customers of, 7, 48, 53, 77–91, 103, 138–39, 147, 151, 173–74, 190–92, 196, 200–201, 225, 231–32, 237
  data networking and, 103–18, 197
  debt burden of, 26
  debt transfer by, 246
  directors of, 37–38, 94, 141, 142–43, 193, 210, 220–24, 248, 250–53, 256–57
  distribution channel stuffing by, 147
  divisions of, 48–50, 81, 96
  downsizing of, 81, 82–84, 260–64

Lucent (*cont.*)
  DSOs and, 150–51
  DWDM systems and, 165–67
  employees' investments in, 261–64
  employees of, 1, 36, 57–59, 62,
    81–84, 93, 95, 121, 128–29, 137,
    197–98, 221, 227, 260–65
  entrepreneurship in, 71, 77, 117
  executive compensation and, 54,
    57–58, 94, 122–23
  fiber optics business of, 53–54
  financial performance of, 53, 147–63
  financial troubles of, 133–77, 180,
    194–99, 222
  forecasting problems of, 201–3,
    213–20, 239, 254
  fraudulent conduct and, 189, 212,
    230
  fund managers and, 197, 243
  global business of, 81–82, 89, 117
  Global Crossing and, 158
  growth prospects of, 52, 53, 77–91,
    97, 122–23, 131–32, 133, 136,
    142–43, 146, 170–73, 180–82,
    200–201, 212
  growth strategy of, 202
  ill-fated Intel deal and, 199–200
  improper transactions of, 233–36
  information systems problems of,
    201–2
  initial planning for, 26–27, 32–36
  Integrated Network Solutions
    division of, 255
  Internet and, 79–80, 136
  inventory build-up of, 83–84,
    151–53, 174, 202–3
  investment bankers and, 135–36, 244
  IPO of, 5–6, 40, 50–59, 64, 135, 154,
    171
  job losses in, 5, 237–40, 260–65

  lawsuits and, 179, 210–19, 257–59
  leadership transfer in, 1–4, 64–65
  McGinn's vision for, 21
  macroeconomic environment and,
    184–87
  management of, 47–48, 62, 64,
    82–83, 90–91, 96–97, 103–4, 107,
    116–18, 121–22, 137–38, 141–42,
    148, 150, 157, 167, 189, 203,
    204–5, 216, 252, 254, 257
  manufacturing outsourced by, 194
  manufacturing problems of, 170–71,
    200–201, 213
  market capitalization of, 5–6, 105,
    179–82
  market forces and, 5–6, 77–91
  merger explorations of, 249–53
  Merrimack Valley Works plant of,
    260–64
  Microelectronics division of, 81,
    107, 150, 151, 168, 198–200,
    211–12, 217, 229, 238
  mission and values of, 33–36, 110,
    121–22, 130–31, 133, 229
  morale collapse in, 227
  naming and branding of, 6, 39–46,
    56, 60–61, 62, 69–70
  North American Service Provider
    Networks division of, 187–92,
    207–19
  optical business of, 201, 203–4,
    205–6, 218, 238
  Optical Fiber division of, 238, 245,
    250, 253
  optical networking acquisition of,
    118–19
  optical technology opportunities
    missed by, 136–37, 153, 164–71,
    180–81, 196, 197, 200–201, 216,
    227

origin and early years of, 5, 9–19, 77
outsourcing and, 194, 262–65
overstaffing of, 239–40
Packet Software of, 172
pension funds of, 175–77
phone centers of, 175
power systems division of, 238
price discounting of, 147–48
product development at, 89, 101
profitability of, 90–91, 93, 122,
    131–32, 133, 149–50, 217
proposed Alcatel merger with, 248,
    249–53
receivables of, 150–51, 174
reorganization of, 96–97, 138–39,
    146, 205
research capabilities of, 69–75
restructuring charges of, 237
restructuring reserve of, 175
revenue of, 31, 56, 70, 78, 80, 93,
    122, 131–32, 133–34, 146,
    148–50, 159, 170–74, 189, 197,
    211, 212, 217, 234, 240
sales contest at, 206–7
sales force of, 110, 111, 112–13,
    121, 123, 131, 133–34, 138–39,
    160, 173–74, 187–92, 202,
    205–6
sales practices of, 146–47, 152,
    155–57, 173–74, 189–92, 202–3,
    207, 217, 233–36
SEC and, 147, 174–77, 189, 212,
    230, 234, 242
shareholders of, 137, 183, 217, 231,
    259–64
spin-offs of, 194–99, 209, 240
Sprint PCS deal with, 79, 154–55
stock analysts and, 70–71, 94,
    122–32, 134–36, 149–50, 155,
    158, 159, 167–68, 170–71,

    173–74, 187–88, 197–98, 214–15,
    225, 251, 256
stock market bubble and, 6–7,
    123–32, 136, 212
stock options and, 57–58, 138
stock price of, 2, 6, 27, 39, 52–53,
    57, 78, 93, 98, 118, 121–32, 133,
    142–43, 171, 173, 179–82, 191,
    194–95, 215, 225, 236, 238, 247,
    254, 255
taxes and, 97–98
Telecommunications Act and, 7
2001 saleable assets of, 238
vendor financing and, 153–64, 212
as Wall Street darling, 2, 77–78,
    90–91, 93–94, 149
weaknesses in, 63–64
Winstar deal with, 160–64, 233–34
wireless business and, 79, 197–98
see also Bell Labs (Lucent);
    Network Systems (Lucent)
Lucent Venture Partners, 119
Lucky, Robert, 67
Lusk, James, 55–56, 59, 146, 148, 154,
    174, 224

Ma Bell, *see* AT&T
McGinn, Rich, 1–5, 13, 15, 19, 21–38,
    39, 41–42, 44, 47, 54, 57–58, 61,
    77, 81, 93, 136, 232, 234–35, 246,
    257, 262
acquisitions and, 97–121, 194–95
appointed President and COO,
    30–31
background of, 21–26
Bell Labs and, 69–75
class action suit against Lucent
    and, 93
company transformation and, 94–97
competitiveness of, 22, 134, 182–83

McGinn, Rich (*cont.*)
  cost structure and, 83–84
  deadpan humor of, 3–4
  Dickson's advice to, 152, 203–4
  directors and, 94
  Ferrari of, 206–7
  Fiorina's departure and, 140–41
  firing of, 4, 220–24
  forecasting and, 202, 247
  global business and, 81–82
  Hopkins and, 192–94, 248
  leadership of, 209–10, 221
  Lucent accounting manipulations
    under, 98–99, 126–32, 147–77
  Lucent's culture and, 62–63, 133
  Lucent spin-offs and, 198–200
  named as CEO, 1–7
  optical technology blunders and,
    167–70, 180–81
  optimism of, 210–11, 218–19
  O'Shea's advice to, 204–5
  passed over as CEO, 30–31
  praise for, 93–94
  removal of Russo and, 207–11
  reorganization and, 96–97, 137–38,
    146, 200
  replacement of Aversano by, 207–19
  Schacht's succession deal with, 3,
    31–32, 64–65
  search for president by, 220
  shareholders and, 217
  sports and, 22
  stock analysts and, 173, 180–82
  stock price and, 98, 194–95
  success of, 136
  vision for Lucent of, 21, 77, 129
  Wall Street and, 226
McGowan, William, 12
Major, Brian, 260
Mandl, Alex, 15–16, 61–62

Marconi, 249
Martin, William McChesney, 184
Marx, William, 15–16, 28
Maryland, University of, 101
Massachusetts Institute of Technology
    (MIT), 66, 67–68
Mathan, Sam, 115
MCI (Microwave Communications of
    America, Inc.), 12, 14, 85–86,
    139, 154, 196
  rejection of OC-192 by, 168
MCI/WorldCom, 241–42
Mellon Bank, 244–45
Mercer Delta, 103, 196–97
Merrimack Valley Works, 260–64
Metropolitan Museum of Art, 29
MFS Communications, 86, 156
Microsoft, 11, 94, 163, 247
money market funds, 242–43
Morgan Stanley, 52–53, 56
Motorola, 27, 126, 131–32, 249
Mumford, Greg, 168
Murray, Cherry, 75
Murray Hill, N.J., 1–4, 65, 66–67,
    69–70, 72, 130, 135, 141, 164,
    194, 214, 250–51, 264–65
mutual funds, 183

Nacchio, Joe, 143–44, 168, 170
Nadler, David, 23, 30, 196–97
NASDAQ, stock market bubble and,
    183
Navy, U.S., 100
NCR (National Cash Register), 13, 104
Nestlé, 63–64
Netravelli, Arun, 199–200
Netscape, 50–51
Network Systems (AT&T), 14, 23, 35,
    47–48, 94–95, 240, 254
  conflicts of interest in, 15

employees of, 47
market domination of, 47–48
optical switches for AT&T and, 89
spin-off of, 5–19, 26–28, 44–48
see also Western Electric
Network Systems (Lucent):
management talent in, 53, 198, 254
reorganization of, 96–97, 138–39
new economy, 6, 122–32, 144,
185–86
New Jersey Conscientious Employee
Protection Act, 215
Newland, J. R., 261
New York *Daily News,* 22
New York Stock Exchange, 58–59
New York Times, 32, 235
Nexabit, 107
Nilsson, Gary, 261–64
*Nina Aversano v. Lucent Technologies,
Inc.,* 210–19
Nobel Prize, 56, 65–66, 68–69, 71
Nokia, 81–82, 249
Nortel Networks (Northern Telecom),
237, 246
ATM sales and, 118
Bay Networks acquisition by, 105,
128
as commercializer of Bell Labs
innovations, 11
as duopoly with Lucent, 94
growth forecasts of, 236–37
as Lucent competitor, 16, 33, 60,
131–32, 136, 167–71, 218
OC-192 systems and, 167–71, 201,
227
optical market and, 136, 167–71,
181, 201, 207, 218, 236–37
optical switching market and, 119
reinvention of, 95
Sprint PCS contract and, 154

stock price of, 128, 194–95, 218–19
vendor financing and, 158–59
NYNEX, 12

OC-3 systems, 167
OC-48 optical system, 166–68
OC-192 transmission system, 164–71,
196, 200, 201, 206, 216, 227, 236
Ocelot, 73–74
O'Neill, Paul, 37–38, 222, 223
optical fiber, 165
optical switches, 89
optical transport, 138, 164–71, 186
optoelectronic components, 50, 136,
153, 199, 247
O'Shea, William, 99, 104, 129–30,
199–200, 202, 204–5, 208, 225,
227, 250

Pacific Telesis Group, 12
Pacino, Al, 225
Packet Software, 172
Packetstar IP Switch, 107
packet switching, 67, 89
Parsippany Hilton, 188, 229, 233
pension plans, 128–29, 175–77
Penzias, Arno, 9, 68–69, 74
Perkins, Donald, 37–38
Perkins, Nellie, 58–59
Peterson, Don, 172–73
Petruschika, Orni, 118–21
Philips Consumer Communications, 49
Philips Electronics NV, 49
photonics, 73
Pirelli Spa, 245
Plaza Hotel, 57
Plunkett, Dan, 103, 141, 256
Porter, Michael, 13
price/earnings (P/E) multiples, 98–99,
128, 130–31, 142

Princess phone, 9
private branch exchange (PBX)
    systems, 48
Prodigy, 80
productivity, 144–45, 185–86
"pull-ups," 203

Qwest, 144, 168–70

radar, 65
Ramqvist, Lars, 15–16
Randall, Rod, 139
Rawson, Richard, 215, 221
Regional Bell Operating Companies
    (RBOCs), *see* Baby Bells
retirement accounts, 183
Rexroad, Brad, 174–75, 182
RJR Nabisco, 63–64
Roth, John, 128, 168, 204
Rouhana, William, Jr., 161, 164,
    233–34
Russo, Pat, 47, 48, 64, 103–4, 116, 129,
    136, 171–72, 197–98, 207–11,
    232, 264
    appointed CEO, 256–57

Sahara Networks, 105
Saudi Arabia, 174, 181, 196, 220,
    236
Savoy Hotel, 54
Schacht, Henry, 1–4, 7, 21–38, 39, 41,
    44, 47, 54–56, 57–58, 61, 93, 136,
    209, 236–40
    accounting practices and, 127, 147
    address to executives by, 228–33
    appointed CEO, 2, 225
    background of, 29
    Bell Labs and, 69–75, 228
    business wisdom of, 7, 162
    CEO search and, 255–57

CLEC servicing and, 89
    deal with Cisco and, 130
    downsizing and, 82–83
    Fiorina's departure and, 140
    firing of McGinn and, 4, 93, 221–24
    forecasting and, 201–3
    growth prospects and, 143
    Hopkins and, 248
    information systems problems and,
        201–2, 226–27
    integrity of, 31, 126
    lawsuits and, 216, 257–59
    leadership qualities of, 29
    Lucent's credibility restored by, 226
    Lucent's culture and, 62–63
    as McGinn's mentor, 64, 65
    McGinn's succession deal with, 3,
        31–32, 64–65
    management style of, 33, 129,
        228–33
    optical technology and, 169–70
    pedantry of, 32–33
    proposed Alcatel merger and,
        250–53
    reorganization and, 139
    restating of revenues by, 235–36
    retirement from chairmanship by, 94,
        257
    sale of fiber business and, 253
    sales contest and, 206–7
    sales practices and, 173–74, 230–31
    stock price and, 98, 226
    telecom industry inexperience of,
        30–31, 54
    thoughtfulness of, 29
    vendor financing and, 159, 160
Schweig, Marc, 31, 43, 110
Securities and Exchange Commission
    (SEC), 40, 123–26, 147, 155,
    160–61, 215, 232, 241–42

Seese, Larry, 168
Senate, U.S., 189
shareholder value, 124–32
Siemens, 50, 81–82, 249
Silicon Valley, 106, 109, 129
Smith, Dan, 137–38
Sony, 11, 165
Southwestern Bell, 12
Sprint, 14, 79, 85–86, 129–30
Standard & Poor's, 176, 243
Stanford University, 51
Stanzione, Dan, 112, 255–56
STM-1, 146
stock analysts, 70–71, 77–78,
     93–94, 109, 123–32, 134–36,
     155, 158, 159, 170–71, 173, 174,
     197–98, 214–15, 226, 241, 251,
     256
stock market bubble (1990s), 5–7,
     124–32, 136, 143–45, 212,
     240–41
stock options, 57–58, 107–8, 114,
     122–23, 129
Stormer, Horst, 71
Stuck, Bart, 87
Sun Microsystems, 11, 22
SUPERCOMM, 107
Sycamore Networks, 105, 137–38
Symons, Jeanette, 109, 111,
     115–16
Szelag, Kathy, 170–71

Tchuruk, Serge, 250–53
Telcordia, 67, 68
TeleChoice, 42
telecom bubble, 156–57
Telecommunications Act (1996),
     84–85, 90
   AT&T and, 16–17
   Lucent and, 7

telecommunications industry, 137
   bankruptcies and scandals in,
      84–85
   barriers to entry in, 80
   changes in, 53, 60, 77–91, 94–95
   collapse of, 5, 145, 185–87, 202,
      219, 228, 236–54
   equipment compatibility in, 157
   Federal Reserve and, 186
   financing of, 7
   growth in, 60, 122, 136, 142–43,
      212
   industry analysts and, 70–71, 77–78,
      86–87
   investment boom in, 6, 77–91, 143
   mania in, 143–45
   outpacing of economy by, 184
   overcapacity in, 145–46
   price war in, 145–46
   spending cuts in, 240
   stabilizing of, 259–60
   stock market bubble and, 5–7
   vendor financing in, 153–64
Telegeography Inc., 145
telephones, 9, 17
   patenting of, 10
Teleport Communications Group
      (TCG), 85–86
Teligent, 62
Telstar satellite, 65
10G (OC-192 optical switch), 169
Theodosopoulos, Nikos, 167
Thomas, Franklin, 37–38, 221,
      223–24
Time, 28, 93
Tomlin, Lily, 17–18
transistors, 65
Trickey, Howard, 74
Truax, Dawn, 80–81, 188
Tyco, 250

Union Pacific, 38
UNIX operating system, 11, 65
U.S. West, 12

Vadasz, Leslie, 199–200
vendor financing, 153–64
    CLECs and, 156–64
    salespeople and, 155–57
venture capitalists, 88, 101, 102, 117,
        119, 143, 156
Verizon, 81, 103
    Network Services Group of, 257
Verwaayen, Ben, 134, 146, 172,
        182–83, 208, 245, 249, 255

Wallenberg family, 16
Wall Street Journal, 49, 127, 186, 235,
        249
Warburg Dillon Reed, 167
Weingarten, Michael, 87
Western Electric, 9–12, 14
    as AT&T's captive monopoly, 10
    government regulation of, 9–10

Merrimack Valley Works plant of,
        260–64
    renaming of, 10, 12
    see also Network Systems (AT&T)
Western Union, 10
Williams, Jeffrey, 52–53, 57
Winstar, 160–64, 202, 233–34
wireless communications and carriers,
        73–74, 79
    see also cellular telephony
WorldCom, purchase of MFS
        Communications by, 86, 154
    see also MCI/WorldCom
Worth, 93

Xerox, 38

Y2K, 78–79
Yale University, 29
Young, John, 37–38, 221
Yurie Systems, 99–104, 141–42, 257

Zhone Technologies, 115–16